TECHNOLOGY SERIES   Technical Memorandum No. 6

Small-scale brickmaking

Prepared under the joint auspices of the International Labour Office and the United Nations Industrial Development Organisation

International Labour Office   Geneva

Copyright © International Labour Organisation 1984

Publications of the International Labour Office enjoy copyright under Protocol 2 of the Universal Copyright Convention. Nevertheless, short excerpts from them may be reproduced without authorisation, on condition that the source is indicated. For rights of reproduction or translation, application should be made to the Publications Branch (Rights and Permissions), International Labour Office, CH-1211 Geneva 22, Switzerland. The International Labour Office welcomes such applications.

ISBN 92-2-103567-0
ISSN 0252-2004

First published 1984
Third impression 1990

The designations employed in ILO publications, which are in conformity with United Nations practice, and the presentation of material therein do not imply the expression of any opinion whatsoever on the part of the International Labour Office concerning the legal status of any country, area or territory or of its authorities, or concerning the delimitation of its frontiers. The responsibility for opinions expressed in signed articles, studies and other contributions rests solely with their authors, and publication does not constitute an endorsement by the International Labour Office of the opinions expressed in them. Reference to names of firms and commercial products and processes does not imply their endorsement by the International Labour Office, and any failure to mention a particular firm, commercial product or process is not a sign of disapproval.

ILO publications can be obtained through major booksellers or ILO local offices in many countries, or direct from ILO Publications, International Labour Office, CH-1211 Geneva 22, Switzerland. A catalogue or list of new publications will be sent free of charge from the above address.

TABLE OF CONTENTS
=================

ACKNOWLEDGEMENTS.. vi

PREFACE.. vii

| CHAPTER I | INTRODUCTION.. | 1 |

	I.	Purpose and objectives of the memorandum	1
	II.	Target audience	3
	III.	Comparison between bricks and other building materials	4
	IV.	Scales of production covered by this memorandum	9
	V.	Content of the memorandum	11

| CHAPTER II | RAW MATERIALS | 13 |

	I.	Origin and distribution of raw materials	13
	II.	Types of clay	16
	III.	Clay testing and significance of results	24

| CHAPTER III | QUARRYING TECHNIQUES | 35 |

	I.	Organisation and management of the quarry	35
	II.	Methods of winning the clay	38
	III.	Transportation to the works	41

| CHAPTER IV | CLAY PREPARATION | 43 |

	I.	Main clay preparation phases	45
	II.	Sorting	45
	III.	Crushing	47
	IV.	Sieving	51
	V.	Proportioning	53
	VI.	Mixing, wetting and tempering	53
	VII.	Testing	58

CHAPTER V	SHAPING		
I.	Description of bricks to be produced		61
II.	Methods of shaping		64
III.	Transportation of bricks to drying areas		79
IV.	Skill requirements and training		79
V.	Productivity of labour		82
CHAPTER VI	DRYING		85
I.	Objectives of drying		85
II.	Artificial drying		86
III.	Natural drying		88
IV.	Shrinkage		95
CHAPTER VII	FIRING		99
I.	Objectives of firing		99
II.	Techniques of firing		100
III.	Kiln design		101
IV.	Auxiliary equipement		134
V.	Fuel		136
VI.	Productivity		139
VII.	Brick testing		141
CHAPTER VIII	MORTARS AND RENDERINGS		149
I.	Purpose and principles		149
II.	Mortar types		150
III.	Mixing and use		154
CHAPTER IX	ORGANISATION OF PRODUCTION		159
I.	Preliminary investigations		159
II.	Infrastructure		160
III.	Layout		163
IV.	Skill requirements		167

CHAPTER X	METHODOLOGICAL FRAMEWORK FOR THE ESTIMATION OF UNIT PRODUCTION COSTS	169
I.	The methodological framework	169
II.	Application of the methodological framework	172
CHAPTER XI	SOCIO-ECONOMIC IMPACT OF ALTERNATIVE BRICK MANUFACTURING TECHNIQUES	177
I.	Employment generation	177
II.	Total investment costs and foreign exchange savings	178
III.	Unit production cost	179
IV.	Rural industrialisation	180
V.	Multiplier effects	181
VI.	Energy requirements	181
VII.	Conclusion	182

APPENDICES

APPENDIX I	Glossary of technical terms	185
APPENDIX II	Bibliographical references	195
APPENDIX III	Institutes from where information can be obtained	201
APPENDIX IV	List of equipment suppliers	207

ACKNOWLEDGEMENTS

The publication of this memorandum has been made possible by a grant from the Swedish International Development Authority (SIDA). The International Labour Office and the United Nations Industrial Development Organisation acknowledge this generous support.

PREFACE

This technical memorandum on small-scale brickmaking is the sixth of a series of memoranda being currently prepared by the ILO and UNIDO. It is the first of three technical memoranda on building materials for low-cost housing.[1]

The main objective of technical memoranda is to provide small-scale producers in developing countries with detailed technical information on small-scale technologies, which have been successfully applied in a number of countries, but are not well known outside the latter. A secondary objective is to assist public planners identify and promote technologies consonant with national socio-economic objectives, such as employment generation, foreign exchange savings, rural industrialisation, or the fulfilment of the basic needs of low-income groups.

The information contained in the memoranda is detailed enough to ensure that small-scale producers should be able, in a large number of cases, to identify and apply the technologies described in the memoranda without the need for further information. Thus, detailed drawings of equipment, which may be manufactured locally, are provided, while a list of equipment suppliers, from both developed and developing countries, may be used for the acquisition of equipment which must be imported. In the few instances where the available information is not sufficient, the reader may obtain additional technical details from publications listed in the bibliography or from technology institutions identified in a separate appendix of the memoranda.

Technical memoranda are not intended as training manuals. It is assumed that the potential users of the technologies described in the memoranda are trained practitioners and that the memoranda are only supposed to provide them with information on alternative technological choices. Memoranda may, however, be used as complementary training material by training institutions.

---

[1] The other two technical memoranda, currently under preparation, cover respectively the production of stabilised earth blocks and that of windows and doors.

This technical memorandum on small-scale brick manufacturing is of particular importance to developing countries as low-cost housing constitutes one of the most important basic needs of low-income groups, and bricks are particularly suitable materials for the construction of this type of housing. Furthermore, the adoption of small-scale brickmaking techniques should generate substantial employment, especially in rural areas. It is hoped that the information contained in this memorandum will slow down the adoption of large-scale, capital-intensive, turnkey brickmaking plants which have often proved to be unsuitable for conditions prevailing in the majority of developing countries.

This memorandum contains 11 chapters, nine of these dealing with the various subprocesses in brick manufacturing. Chapter X provides a methodological framework for the estimation of the unit production cost of bricks, using the technical data from the previous chapters. It is of particular interest to potential brickmakers who wish to identify the least-cost or most profitable production technique. Chapter XI is mostly intended for public planners and project evaluators from industrial development agencies who wish to obtain information on the various socio-economic effects of alternative brickmaking techniques with a view to identifying and promoting those which are particularly suitable to local socio-economic conditions.

This memorandum also contains four appendices which could be of interest to the reader. Appendix I provides a glossary of technical terms, and should therefore be of assistance to non-specialists. Appendices II and III provide sources of additional information, either from available publications (Appendix II) or from specialised technology institutions (Appendix III). Finally, Appendix IV provides a list of equipment suppliers from both developing and developed countries. It may be noted that this list is far from being exhaustive, and that it does not imply a special endorsement of these suppliers by the ILO. The listed names are only provided for illustrative purposes, and brickmakers should try to obtain information from as many suppliers as feasible.

A questionnaire is attached at the end of the memorandum for those readers who may wish to send to the ILO or UNIDO their comments and observations on the content and usefulness of this publication. These will be taken into consideration in the future preparation of additional technical memoranda.

This memorandum was prepared by Mr. R.G. Smith (consultant) in collaboration with Mr. M. Allal, staff member in charge of the preparation of the Technical Memoranda series within the Technology and Employment Branch of the International Labour Office.

A. S. Bhalla,
Chief,
Technology and Employment Branch.

CHAPTER I

INTRODUCTION

I. PURPOSE AND OBJECTIVES OF THE MEMORANDUM

Housing constitutes one of the most important basic needs of low-income groups in developing countries. However, it is the most difficult to satisfy as land and building costs are often outside the means of the unemployed and underemployed in both rural and urban areas. Thus, many governments have launched various schemes with a view to facilitating housing ownership by low-income groups, including self-help housing schemes, granting of housing subsidies, provision of credit at low interest rates, etc. Given the limited means at the disposal of governments and potential home owners, it is important to seek ways to lower the cost of low-income housing while minimising repair and maintenance costs. In particular, governments should promote the production and use of cheap yet durable building materials as the latter constitute a very large proportion of total low-income housing costs in developing countries(1). Furthermore, it would be useful if the production of these building materials could contribute to the fulfilment of important development objectives of these countries, such as the generation of productive employment, rural industrialisation, and a decreased dependence on essential imports.

A number of traditional building materials exist which have proved themselves to be the most suitable materials for use in a wide variety of situations, and have a great potential for increased use in the future. These traditional materials, which make use of locally available raw materials, can be manufactured close to the construction site with little equipment (which may be produced locally), and are often more appropriate to the environment than modern materials. One such building material is clay bricks. The purpose of this technical memorandum is to provide detailed technical and economic information on small-scale brick manufacturing with a view to assisting rural and urban entrepreneurs to start up new plants or improving their production techniques. It is also hoped that the information contained in this memorandum will help slow down the establishment of large-scale, capital-intensive plants which are not always suitable to socio-economic conditions of developing countries.

I.1 Need for improved brick production techniques

Various production methods are used for brickmaking in developing countries. Traditional hand digging, moulding and handling are used by a large number of small production units. Some larger units tend to use equipment for digging or mixing, while a number of developing countries have chosen to import large-scale capital-intensive plants.

The choice of brickmaking technology is mostly a function of market demand (e.g. scale and location of demand, required quality standard), availability of investment funds, and unit production costs associated with alternative production techniques. In some cases, governments may also impose various policy measures with a view to favouring the adoption of techniques consonant with the national development objectives. Whatever the adopted technique, quality may be improved and costs reduced if appropriate measures are taken during the production process.

Experience shows that a large fraction of bricks are often wasted during the various production stages. For example, moulded bricks get eroded by the rain before firing or distorted by bad handling methods. Sometimes, incorrectly adjusted machines yield inconsistent or inferior output which may not be marketed. With attention to the basic principles of brickmaking and more care, a greater number of bricks could be produced for the same expenditure of labour, raw materials and fuel.

In some instances, more careful preparation of raw materials would minimise problems at subsequent manufacturing stages. For example, if stones or hard dry lumps of clay are included in the moist clay used for moulding, they will exhibit different drying shrinkages from the moist clay and give rise to cracks in the dried or fired bricks. The remedy in such a case would have been to select a more uniform raw material, or to remove the offending particles, or to break the material down to a fine size (e.g. by manual means or with a crushing machine).

Use of a good product, of regular shape and size and of consistent properties, will enable the accurate building of walls while minimising the use of mortar between bricks. Renderings, often applied in developing countries, will also require less mix for a given wall area if the brickwork face is accurate. Alternatively, if the brick quality is sufficiently good, it may be unnecessary to apply renderings at all. A good product will thus favourably impress the customer and save materials, time and money. It should therefore improve future demand for the product. Good bricks should be durable and brickwork should be long lasting.

I.2 Availability of information on brickmaking

The techniques of brickmaking are often handed down from father to son in small works, or are taught in various technical schools, training centres, etc. Articles and books have been published(2) but are often too brief or mostly concerned with large-scale production, scientific investigations or laboratory tests. They also often relate to conditions and needs of the more developed countries. With few exceptions(3), there is a lack of information on practical details of small-scale production in rural or peri-urban areas.

Information on sophisticated high capital cost brickmaking plants can be obtained from published books and scientific and trade journals, or from equipment manufacturers and consultants. On the other hand, it is more difficult to obtain information on small-scale, labour-intensive production. Many appropriate technology institutes, building research centres and university departments do generate information on appropriate production techniques (see list in Appendix III). However, this information is either not published or is not disseminated to other developing countries. This memorandum seeks, therefore, to provide information on small-scale brickmaking with a view to partially filling the current information gap. It does not provide technical details on all possible circumstances, but will, it may be hoped, induce small-scale producers to try production techniques which have already been successfully adopted in a number of developing countries.

II. TARGET AUDIENCE

This memorandum is intended for several groups of individuals in developing countries, including the following:

- small-scale brickmaking producers in rural and urban areas, and those considering starting brick production. These could be either individual entrepreneurs or groups of artisans associated in a manufacturing co-operative. These producers will be mostly interested in the information contained in Chapters II to VII, and Chapters IX and X.
- housing authorities, public planners and project evaluators in various industrial development agencies may be interested in the information contained in Chapters II and XI. This latter chapter, which focuses on the socio-economic implications of alternative brickmaking techniques, will be of particular interest to public planners concerned with employment generation, foreign exchange saving, etc.
- financial institutions, businessmen, government officials and banks should

be mostly interested in Chapter X which provides the necessary information for costing alternative production techniques.

- handicraft promotion institutions, village crafts organisations and equipment manufacturers should find the technical chapters II to VII useful.

- voluntary organisations, foreign experts, extension workers and staff of technical colleges will wish to compare bricks with other building materials, as detailed in Chapter I (section III). They may also benefit from the technical information contained in Chapters II to VII.

It must be stressed that this is not a technical memorandum on the use of bricks in building, although some of the information contained in Chapter VIII may be of interest to builders.

III. COMPARISON BETWEEN BRICKS AND OTHER BUILDING MATERIALS

This section compares the properties of fired clay bricks with those of other alternative walling materials. Table I.1 gives specific values for some of the properties discussed below. This comparison should be useful to housing authorities in deciding which building materials should be most appropriate for various types of housing or housing projects.

III.1. Strength

Compressive strength of fired-clay bricks varies enormously, depending upon clay type and processing. Strength requirements for single-storey housing are easily met.

Calcium silicate bricks, made from sand with high silica content and good quality, low magnesia, lime, may have strengths approaching those of good fired clay bricks. However, high capital cost machinery is used for the mixing, pressing and autoclaving. Furthermore, calcium silicate bricks must be produced in large-scale plants.

Concrete bricks and blocks have sufficient strength, but require cement which is expensive, and must often be imported.

Lightweight concrete blocks, made with either natural or artificial lightweight aggregate, have adequate strength but require cement.

Aerated concrete has low strength which may be sufficient for one-storey buildings. Particularly careful production control is necessary, using autoclaving to reduce subsequent moisture shrinkage of blocks made from this material.

Many types of soil have sufficient compressive strength when dry. However, this strength is considerably reduced once they become saturated with water.

Table I.1

Range of properties of bricks and blocks

Property	Fired clay bricks	Calcium silicate bricks	Dense concrete bricks	Aerated concrete blocks	Lightweight concrete blocks	Stabilised soil blocks
Wet compressive strength (MN/m^2)	10 to 60	10 to 55	7 to 50	2 to 6	2 to 20	1 to 40
Reversible moisture movement (% linear)	0 to 0.02	0.01 to 0.035	0.02 to 0.05	0.05 to 0.10	0.04 to 0.08	0.02 to 0.2
Density (g/cm^3)	1.4 to 2.4	1.6 to 2.1	1.7 to 2.2	0.4 to 0.9	0.6 to 1.6	1.5 to 1.9
Thermal conductivity ($W/m\,^0C$)	0.7 to 1.3	1.1 to 1.6	1.0 to 1.7	0.1 to 0.2	0.15 to 0.7	0.5 to 0.7
Durability under severe natural exposure	Excellent to very poor	Good to moderate	Good to poor	Good to moderate	Good to poor	Good to very poor

Waterproofers such as bitumen, or stabilisers such as lime or cement, may be used with certain soils to improve wet strength. On the other hand, the strength of the other previously mentioned materials is only reduced slightly when they are wetted.

Gypsum, which occurs as a soft rock, or in some places as a fine sand, can be converted to plaster by gentle heating and then mixed with fine and coarse aggregate and cast into building blocks(5). Strength will be adequate for single-storey constructions, though wetting will reduce compressive strength to 50 per cent of the dry value.

Thus bricks are seen to be at the top of the list for strength, especially when wet.

Many other walling systems exist, notably panels made either from woven plant leaves or stems, or manufactured from cement, plastics, wood or metal. However, the strength of walls made from these panels will depend largely upon the frame which is built to hold them.

III.2. Moisture movement

Most porous building materials expand when wetted and contract again as they dry. Excessive movement can cause spalling, cracking or other failures in buildings. This reversible expansion is very small in fired clay bricks. However, a slow irreversible expansion commences as soon as bricks leave the kiln. This irreversible expansion may vary from virtually zero to 0.1 per cent linear movement. Under normal circumstances much of this expansion will have taken place before bricks are built into walls. Thus, the remaining expansion is likely to be insignificant in the context of small buildings(6).

Properly made calcium silicate bricks and concrete bricks are unlikely to have more than a fairly small amount of moisture movement. However, lightweight and aerated concrete units exhibit a greater movement. This sometimes leads to shrinkage cracking in buildings as they dry out initially.

Soil, especially plastic clay, may have a very large moisture movement of several percentage points. This is a major cause of failure in earth building. The problem is reduced if stabilisers are incorporated into the soil.

Timber, bamboo and other plant materials exhibit variable, but sometimes large, moisture movements. The latter take place especially across the grain rather than in line with it.

Moisture movement becomes especially important when two materials with different movement characteristics are in close juxtaposition in a building. Differential movements give rise to stress which may be sufficient to break the bond between the materials, or lead to other damage. For example, cement renderings often become detached from mud walling, and gaps appear sometimes between timber frames and infill materials.

Bricks thus compare favourably with alternative construction materials. Moreover, brickwork can be built without timber frames, thus excluding the possibility for differential movements.

III.3. Density and thermal properties

Fired clay bricks are amongst the most dense of building materials. This high density may constitute a disadvantage for transportation over long distances or in multi-storey framed buildings where the loads on frames would be high. On the one hand, weight is of little consequence whenever bricks are produced locally for close-by markets and single-storey buildings. On the other hand, the high density of bricks has the advantage over lightweight building materials of greater thermal capacity. This characteristic is sought in the tropics where extremes of temperature will be moderated inside buildings made of bricks.

Aerated and lightweight aggregate concretes have good thermal insulating properties but lack thermal capacity, while thick mud walls have fairly good insulation and good thermal capacity. Lightweight cladding materials, such as woven leaves and matting, metal sheeting and asbestos cement sheeting, have neither high insulation nor high thermal capacity.

Thus, bricks are particularly advantageous for low-cost housing as they considerably improve environmental conditions within the building.

III.4. Durability, appearance and maintenance

Evidence for the excellent durability of brickwork may be seen in many countries of the world. In the Middle East, brickwork 4,000 years old still remains. Bricks made 2,000 years ago in Roman times are still in use today. Indeed, properly made bricks are amongst the most durable of materials, having typical properties of ceramics such as good strength, resistance to abrasion, sunlight, heat and water, excellent resistance to chemicals and attacks by insects and bacteria, etc. If bricks are not well made (e.g. if the time or temperature in the kiln is insufficient), these desirable ceramic properties will not be developed, and performance will be nearer to that of mud bricks. Furthermore, to achieve the best performance from brickwork, attention must be paid to the correct formulation and use of mortar. Fired clay brickwork should sustain the adverse impact of the environment without the need of any surface protection (e.g. rendering). No maintenance should be required subsequent to building.

In some communities it is traditional to render the brick wall surface, although this is not necessary from either the appearance or performance point of view. Furthermore, lime washing on rendering is often used to achieve a white finish. A white finish is beneficial in reducing solar gain. It may be applied directly on the bricks without rendering, thus saving materials.

Other brick and block materials may also have good durability, if well made. However, the lightweight aggregate, aerated concrete and mud bricks normally require rendering to improve resistance to water.

The ultra-violet component of sunlight causes deterioration of many organic materials. These include timber and other plant-derived materials, plastics, paints, varnishes and bitumen. Inorganic materials, such as bricks, are immune to sunlight deterioration.

Termites occur in many developing countries and can attack and damage soft materials such as various species of timber. Other insects also attack timber. Hard materials such as bricks are entirely resistant.

Under damp conditions, timber and many other organic materials may rot through attacks by fungi, moulds and bacteria. Although some plant growth and mould may be seen on porous inorganic building materials, especially in hot-damp climates, damage is unlikely.

Fire can quickly destroy many building materials such as timber, woven matting and plastics. Cement products do not burn, but high temperatures in fires could break down some of the calcium and alumino silicates of which they are composed, causing loss of strength. In practice concrete may successfully sustain somewhat elevated temperatures without serious effect, though if concrete contains siliceous aggregates, such as flint, it is likely to spall(7). Reinforcing steel and steel frames lose strength and distort in fire, but are normally encased, and are thus protected to a large extent. Although clay brickwork could spall, crack and bulge in a severe fire, bricks are less likely to suffer damage than concrete and calcium silicate bricks as they have already been exposed to fire. Sudden cooling of hot areas by quenching with water in the course of fire-fighting may cause spalling. This does not generally affect the strength and stability of the brick wall seriously.

III.5. Earthquake areas

In general, bricks and blocks, whether of fired clay, calcium silicate, concrete, or stabilised soil, require steel reinforcement in seismic zones. Mud building, lightweight concrete and aerated concrete materials will also be at risk in these areas if not similarly reinforced.

III.6. Production cost and foreign exchange

The production costs of various building materials depend upon raw materials prices, methods used, markets, etc. which vary from time to time and place to place, making comparisons difficult. However, bricks are often amongst the cheapest of walling materials. It should be borne in mind that if, as stated at a United Nations Conference, a house is to retain its usefulness, it must be maintained, repaired, adapted and renovated. Thus, choices concerning standards, materials and technology should consider resource requirements over the whole expected life of the asset and not merely the monetary cost of its initial production(8). Durable materials such as bricks have a cost advantage in this respect.

The production of bricks from indigenous clays, especially if labour-intensive methods are used to avoid importation of capital-intensive equipment, will conserve foreign exchange. This is in contrast with some of the alternative materials.

IV. SCALES OF PRODUCTION COVERED BY THIS MEMORANDUM

Bricks manufacturing may be undertaken at various scales of production, depending upon local circumstances. Table I.2 summarises the production techniques used at small, medium and large scales of production.

Table I.2

Scales of production in brick manufacturing

Scale of production	Number of bricks per day (average)	Example of process used	Appropriate for market area
Small	1 000	Hand made, clamp-burnt	Rural village
Medium	10 000	Mechanised press, Bull's trench kiln	Near towns
Large	100 000	Fully automated Extruded wire cut, tunnel kiln	Industrialised areas of high demand and well-developed infrastructure

This memorandum is concerned primarily with small-scale production, though some consideration is given to medium-scale. Large-scale will be mentioned briefly for comparative purposes.

IV.1 Small-scale concept

A small brickworks producing 1,000 bricks per day may supply enough bricks each week for the building of an average size house. This may be adequate in a small village community. However, if demand were to increase suddenly, production could be increased to several thousand bricks per day merely by making additional wooden moulds and hiring more workers. In this case, the management staff does not need to be expanded. This larger production unit might also be established in small towns. Conversely, at times of recession or when weather prevents construction work, the demand will fall and production can be reduced temporarily. Thus, such small-scale industry is very adaptable to a changing market.

Small-scale production should be undertaken near the clay source, and within a short distance of the area where bricks will be sold and used. This will reduce transport cost while saving fuel. Small-scale production will not unduly spoil the landscape nor cause excessive pollution. An electricity supply may not be necessary and fossil fuels need not be used. Kilns may utilise waste materials for fuel, such as saw dust, rice husks, animal dung and scrub wood. The small works will provide employment within the local community. Capital investment is low for small-scale production and is thus appropriate for poor communities. Furthermore, equipment for small-scale brickmaking can be made and repaired within the local community.

IV.2. Large-scale concept

In contrast to the points mentioned above, the introduction of large brickworks necessitates capital investment of millions of dollars, mostly in foreign exchange for the import of the sophisticated production machines and control systems. Commissioning over a period of months and subsequent purchase of spares will further increase costs. Large areas will have to be cleared not only for the works, but also for the clay pit. The process itself and the transport of raw materials and products can prove a nuisance. Production of many millions of bricks per year necessitates the finding of sufficient markets, and involves the use of fuel for getting bricks to the building sites. Feasibility studies for large-scale plants commonly assume several shifts being worked per day for nearly all the days of the year. Such plants are not adaptable to variations in market demand. There is no allowance for workers' absenteeism (e.g. during the agricultural season), neither do they ususally take into account the difficulty of obtaining spare parts from overseas in the event of a breakdown. If the latter does occur, the whole production ceases. These large plants require electric power and high grades of fuel for the kilns.

In those situations where a large plant may be considered, it would be normal to conduct a full feasibility study, examining raw material quality and reserves for the expected life of the works (e.g. 50 years), and a thorough market survey. A specialist consultant would be required for such a feasibility study.

V. CONTENT OF THE MEMORANDUM

The following eight chapters of this memorandum deal with the technical aspects of brick manufacturing. Chapter II describes various raw materials entering in the production of bricks while Chapters III to IX describe the various production processes in the following order:

Chapter III   : Quarrying (methods and equipment)
Chapter IV   : Preprocessing (grinding, sieving, wetting, etc.)
Chapter V   : Forming (equipment, skill requirements, etc.)
Chapter VI   : Drying (natural and artificial, drying shrinkage, etc.)
Chapter VII   : Firing (kiln types, fuels, etc.)
Chapter VIII : Mortars and renderings (purpose, types, etc.)
Chapter IX   : Organisation of production (plant layout, water and fuel supplies, labour, etc.)

Technical details on each subprocess are provided, including advice for improving product quality, saving fuel, increasing labour productivity, minimising losses, etc.

Chapter X outlines a methodological framework for estimating unit production costs associated with alternative production techniques. An illustrative example is provided with a view to showing how this framework may be applied to a specific bricks production unit. Finally, Chapter XI analyses the various socio-economic effects of alternative production techniques, including employment generation, foreign exchange savings, fuel utilisation, etc. The memorandum concludes with the following appendices: Glossary of technical terms, bibliography, list of institutions concerned with brickmaking, and list of equipment suppliers.

Note:

[1] The references are to entries in the bibliography (Appendix II).

CHAPTER II

RAW MATERIALS

I. ORIGIN AND DISTRIBUTION OF RAW MATERIALS

Brickmaking requires sufficient supplies of suitable soil, sand, water and fuel. The purpose of this chapter is to describe the various types of clay which may be used in brickmaking.

The essential ingredient in the soil used for brickmaking is clay. The size of each clay particle is extremely small, generally less than 0.002 mm across. Various forces act between these fine particles in a moistened clay, allowing the latter to be formed into the desired shape, which must be retained on drying. Clayey materials can be readily identified by simple manipulation of moist samples with a view to checking the plasticity of the latter.

A wide variety of raw materials may be used for brickmaking, ranging from soft sticky muds to hard shales. However, all these materials must contain a moderate proportion of clay-size particles. Too high a proportion of such particles will result in excessive shrinkage of moulded bricks as they dry, with consequent risk of cracking. On the other hand, a soil with too low a proportion of clay particles will not be cohesive enough and will fall apart. The mineralogical nature of the clay must be suitable so that it is changed by heating in a kiln to a strong, water resistant vitrified form which can bind larger particles in the soil together.

Brickmaking clays may be found in most countries of the world. Geologically recent deposits are associated with existing valleys and rivers, and are often near the surface. Older deposits may be overlaid by other unsuitable material of varying depth, and may have been raised and inclined from their original positions. Thus, good deposits of clay may be found in gently rolling hills, but not mountains.

Information on clay deposits is available in many countries from National Geological Survey Departments, or may be obtained from Geological Institutes.

Location of existing brickworks, pottery works or other ceramic production is evidence of workable deposits.

Prospecting for new clay deposits may be undertaken by first examining river banks, and the sides of any recent road or railway cuttings which give an instant section of the soil profile. Subsequently it is necessary to explore in more detail any newly-discovered deposits by taking samples from many points on a regular grid covering the ground area. The neatest and simplest means of obtaining a suitable sample is by using an earth auger. The latter can be powered by one or two people. As it is rotated, the auger drills its way down into the earth, providing samples of the cut out soil. Alternatively, a spade may be used to dig a narrow hole (figure II.2). However, it cannot go as deep as an auger. A pit may be dug instead in such a manner that a person with the spade can work on the floor of the pit. This will require the removal of a great volume of earth, and may not therefore constitute an efficient way of taking samples. For safety's sake the pit should not be more than 2 metres deep.

It is wise to keep an accurate record of such investigations. A plan of the area should be drawn, and location of investigatory holes marked in and numbered. Samples taken out of the hole should be small enough to allow the identification of a change from one soil type to another. Usually, there is a top-soil in which plants grow, and which contains the decomposed products of plants. The top-soil depth should be measured and noted, as well as that of subsequent soil layers. As soon as clay is found, it will be recognised by the stickiness with which it adheres to the auger or spade. If a large stone is encountered when augering, it will have to be knocked out of the way, or broken, or a different type of auger used to cut a way past.

The survey will indicate the area covered by clay, its thickness and the depth at which it may be found, and the thickness of the top soil which must be removed during quarrying. If there is much top soil, it will not be worth the cost of removing it unless there is a good depth of clay beneath.

Simple testing of clay for suitability for brickmaking may be carried out on site. For more extensive testing, each soil type should be in a separate heap on boards or a large sheet, then reduced by quartering. Quartering is done by dividing the heap into four quarters of equal size and shape, discarding two diagonally-opposed quarters, and recombining the other two. This procedure is repeated until a small pile of a few kilograms remains. The latter should be placed in a strong plastic bag, labelled with the hole numbers and the depths from which the sample was extracted.

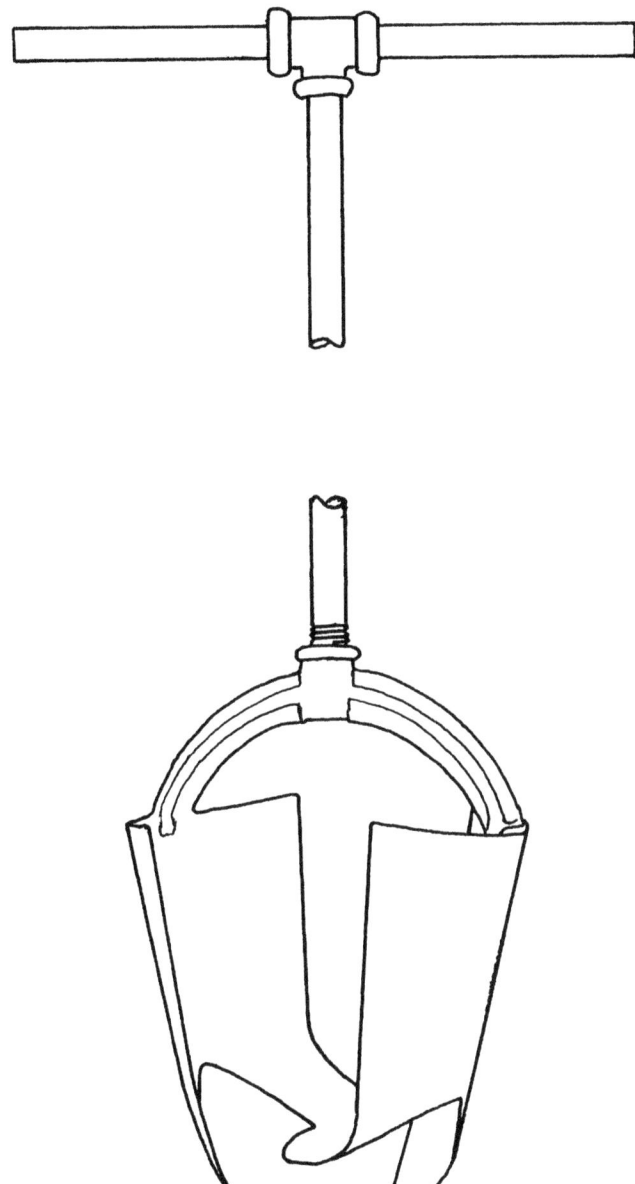

Figure II.1

Bucket auger for sampling soil

II. TYPES OF CLAY

It is essential that the raw material used for the production of bricks contains the following elements:

- sufficient clay fraction to ensure a good plasticity of the clay body, thus allowing the latter to be formed and retain its shape. The material is described as 'lean' or 'short' if the fine fraction is insufficient. The clay element should not exceed a certain limit which will render it too sticky for working. Furthermore, the dried bricks are liable to cracking due to high shrinkage if too much clay is present in the body. In this case, the material is described as 'fat'. Some clay types with the above characteristic have high shrinkage rates;

- sufficient unreactive coarser grained material such as sand to mitigate the potential problem described above;

- proportions of silica and alumina in the clay from which the strong durable glassy material may be formed on heating to approximately $1000^0 C$;

- alkalis or iron to assist in the formation of glassy compounds;

- constituents which do not produce excessive deformation or shrinkage at the firing temperature in the kiln;

- no impurities or inclusions which will disrupt the structure of the brick.

The size of particle present in the clay body affects the cohesiveness, forming characteristics, drying and firing properties of a clay.

II.1 Particles sizes in brickmaking soils

The various fractions of particles in soils are usually denoted by their size as given in Table II.1.

Table II.1

Definition of particle sizes in brickmaking soils

Fraction		Size range (mm)
Sand	Coarse	2 - 0.6
	Medium	0.6 - 0.2
	Fine	0.2 - 0.06
Silt	Coarse	0.06 - 0.02
	Medium	0-02 - 0.006
	Fine	0.006 - 0.002
Clay		less than 0.002

In practice, a raw material for brickmaking should contain some clay fraction (say 10 to 50 per cent) together with some silt and some sand. Depending upon relative proportions of various elements in the raw material, the latter might be described, for example, as a silty clay or, if containing some clay and similar proportions of silt and sand, as a loam. Since the presence of both clay and a good range of other particle sizes is desirable, loams are particularly suitable for brickmaking.

II.2 **Clay minerals**

Materials for brickmaking range from soft muds through the partially compacted clays or muds and highly compressed shales. The fine particles in the clay fraction may consist of various mixtures of some 12 different groups of clay minerals. These groups are briefly described below.

The kaolin group is common and might be regarded as a typical clay mineral. In its molecular structure thousands of alternate flat layers of silica (silicon oxide) and gibbsite (aluminium oxide) occur, and give the particles their typical hexagonal plate-like structure. They are up to 0.002 mm across and can be seen under the electron microscope. This mineral presents no particular problems in brickmaking.

The montmorillonite group, which often occurs in the drier tropics, has two silica layers for every one gibbsite. This structure allows water molecules to enter in between the layers, forcing them apart. The resulting expansion of the clay may continue for several weeks under damp conditions(17) The layers close up again when the water is dried out. This has important consequences in brickmaking since montmorillonite-bearing clays have large drying shrinkages. The thin plates are generally smaller than kaolinite. The high specific surface area gives great plasticity, stickiness and strength to the montmorillonites(17).

The hydrous micas and illites, which have somewhat similar structures to the montmorillonites, are also frequently found in brickmaking materials. Chlorites, which are related to hydrous micas, are also found in various clay materials. The latter have magnesium and potassium within their structures.

Extremely small particles from a millionth to a thousandth of a millimetre across, termed colloids, are also present in clays. They carry electrical charges, so their movement in water and their properties are affected by the presence of salts. Thus, the physical properties of wet clays can be altered by additions of some chemicals which may, for example, increase their plasticity or reduce stickiness. An acidic addition flocculates the colloidal particles so they settle in water more readily whilst an alkaline addition deflocculates these particles and keeps them in suspension.

Mineralogical examination can help identify the substances present with a view to determining the likely suitability of a material for brickmaking.

II.3 Chemical analysis

Chemical analysis can help in the identification of the clay minerals present in the raw material. The relative proportions of silica and alumina are relevant, since the higher the proportion of alumina, the higher the temperature necessary to form the glassy ceramic bonding material which characterises ceramic products. Chemical analysis can also indicate the presence of water-soluble compounds such as the sulphates of potassium, sodium and magnesium. The drying out of the latter on the moulded bricks (before firing) produce unsightly scumming. If still present in the fired product, they may lead to efflorescence and, exceptionally, can spoil brick faces and lead to attack and expansion of cement-based mortars. Calcium sulphate can also produce this undesirable effect. With knowledge of these deleterious salts within the clay it might be possible to avoid problems with the bricks when finally built into walls, by choosing another clay deposit or allowing rain to wash salts out of the clay after it has been dug, or by firing the bricks to a higher temperature. Another solution to these problems is to add barium carbonate. This is, however, an expensive remedy which may not be feasible in many situations.

If potassium and sodium are found in the chemical analysis, but the compounds are not water soluble, they may indicate the presence of fluxes such as the felspars or micas. These are beneficial in reducing the temperature needed for formation of glassy material. Magnesium, calcium and iron (ferrous) compounds can also behave as fluxes.

Chemical analysis may be carried out on different size fractions of the soil. This is an important consideration since fluxes should be in the finest of particles sizes. Hence, their presence in only coarse fractions is of little significance.

Laterites occur as rock, gravel, sand, silt and clay in many tropical locations. They are high in alumina and low in silica. Thus, the use of laterite soils for brickmaking will require higher kiln temperatures. In practice, the presence of potassium and sodium-bearing compounds, and of iron compounds (which are often abundant and act as fluxes), should allow the production of bricks from laterites. The latter are defined in a number of ways, but the following definition is often accepted: The ratio of silica to sesquioxides (that is iron and aluminium oxides) must be less than 1.33 for the material to be a laterite. If the ratio is between 1.33 and 2, the material is lateritic, and if the ratio is greater than 2, it is non-lateritic.

Marls, which are clays with a high proportion of calcium carbonate (chalk, limestone, etc.), are identified by high calcium and high weight losses on heating in a full chemical analysis. They may have low vitrification temperatures which extend over only a narrow range. Thus, sudden fusion can occur in manufacture. If the calcium carbonate is present as large lumps, the latter will have a disruptive effect on the fired bricks after manufacture. These lumps should be removed or ground to less than 2 mm.

II.4 The drying process and drying shrinkage

A wet clay has the fine individual particles separated by films of water which are absorbed into the particle surfaces. In such a state the clay exhibits its typical plastic property which enables it to be shaped. On drying, the films are reduced and the particles get gradually closer. Thus, an overall shrinkage of the body is discernable. The shrinkage continues until the particles touch, but water still remains in voids between the particles. The clay then has a critical moisture content (CMC). As the water continues to dry out, no further significant shrinkage occurs. This is shown diagramatically in figure II.2. The practical significance of the process is that bricks must be dried slowly to the CMC, thus ensuring that all parts of the brick (top, bottom and inside) are shrinking at the same rate. If one face of a brick dries before the opposite face and becomes non-plastic, the latter face may crack as it dries while being held in position by the dried face. Different rates of shrinkage also cause bricks to become bowed, or banana-shaped by a similar process. Once the CMC is reached, faster drying may be used since there is no further shrinkage.

Clays for brickmaking should not have too high a shrinkage rate on drying if cracking is to be avoided. However, if the moulded bricks are dried very slowly, higher shrinkage material may be used. Montmorillonite has an exceptionally large drying shrinkage, so soils containing it (e.g. black cotton soils) would be best moulded from the driest possible mix, and then dried very slowly. In general, the greater the proportion of fine particles the greater the drying shrinkage, and the finer the particles the more the shrinkage. Hence, there should not be too much clay in brickmaking soil.

To reduce unacceptably high shrinkages, non-reactive coarse grained material may be mixed in with the soil. The additional materials frequently employed are sand, if it is available nearby, or ground-up reject bricks which are referred to as 'grog'.

Drying should be as complete as possible before bricks are exposed to the heat of the kiln. Otherwise, steam may be produced in the bricks and develop enough pressure to blow them apart (other reasons are listed in Section I of Chapter VI).

II.5 The firing process and firing shrinkage

At a low temperature of $100^{\circ}C$, any moisture remaining in the bricks is removed. The nature of the clay is not changed (i.e. the cooled and wetted clay retain its original characteristics - see figure II.3).

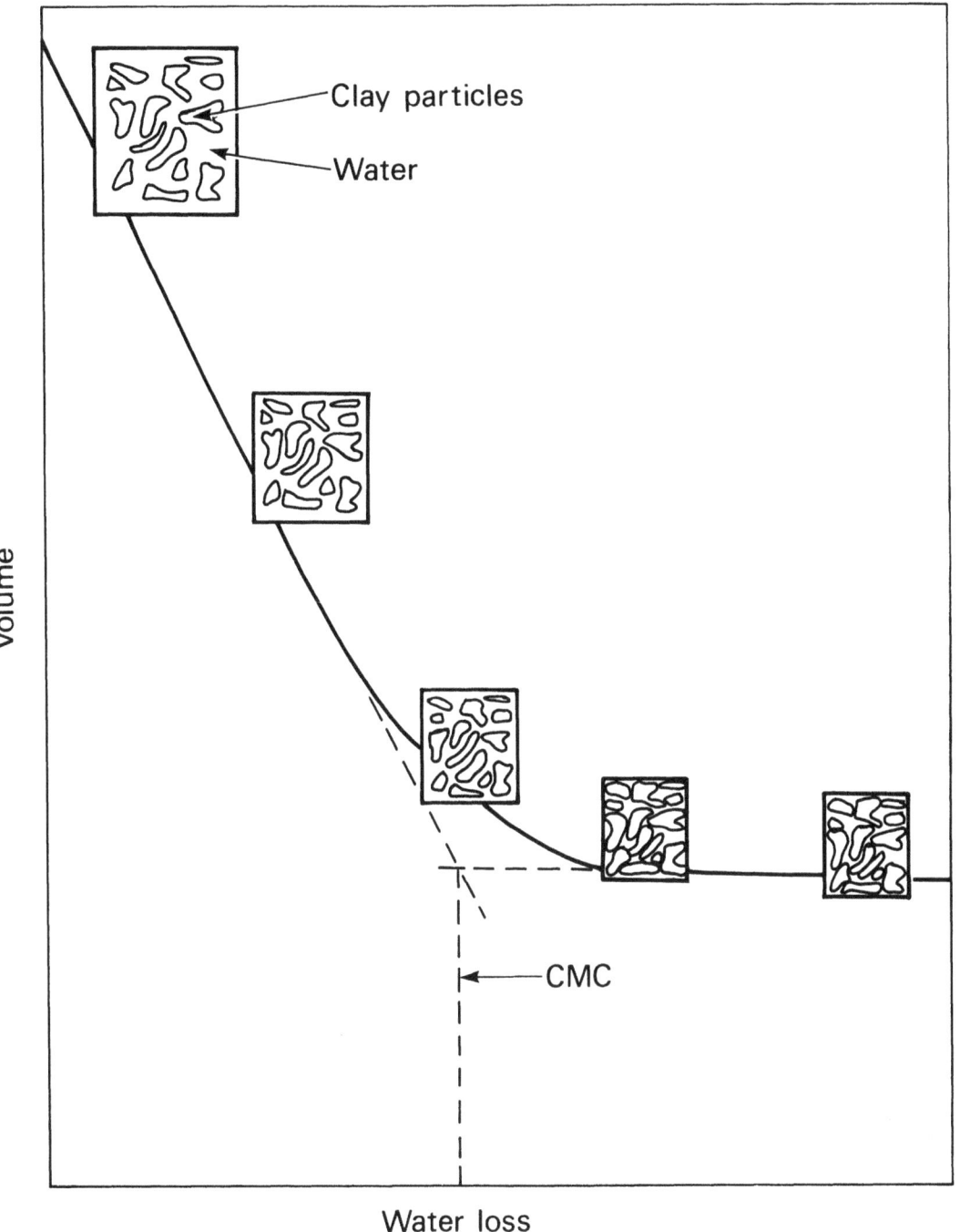

Figure II.2

Drying curve

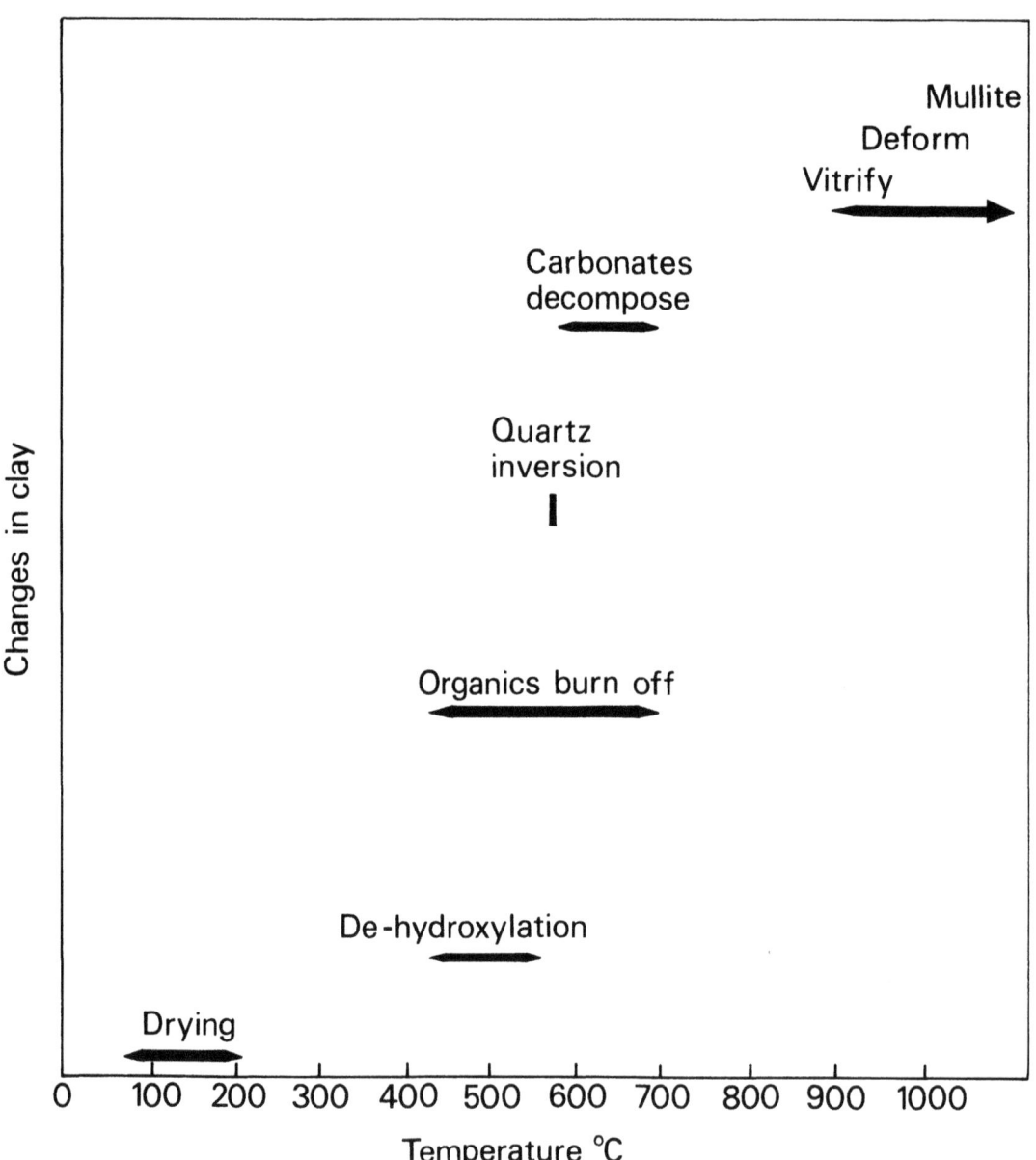

The first irreversible reactions start at approximately 450-500°C, when dehydroxylation takes place. Part of the actual clay structure (the hydroxyl groups) is driven off as steam, resulting in a very small expansion of the brick.

Carbonaceous organic matter (derived from plants, etc.) in the soil will burn off in the temperature range of 400-700°C, provided sufficient air is allowed in to convert it to carbon dioxide gas. Time is required for the brick to heat up, for oxygen to diffuse in, and for carbon dioxide to diffuse out. If this organic matter is not completely burnt off before the temperature rises to the point at which glassy material forms, the diffusion processes will not be possible, and carbon will remain within the bricks as undesirable black cores. The latter may also be caused by the lack of oxygen. An "opening material", such as a burnt refractory clay, can be mixed in to aid gas diffusion.

Present carbonates and sulphides decompose at the top of the temperature range at which the organic matter is burnt, carbon dioxide and sulphur dioxide being given off.

Silica, which is a common constituent of brickmaking soils in the form of quartz, changes its crystal form at 573°C. This so-called inversion is accompanied by an expansion. Consequently, the rate of rise of temperature must be slow if one is to obtain near-uniform temperature throughout the brick and thus avoid excessive stresses which could lead to cracking.

The glass formation, which is necessary to bond particles together and make the product strong and durable, commences at approximately 900°C, depending upon the composition of the soil used. The process, known as vitrification, involves fluxes reacting with the various other minerals in the soil to form a liquid. The higher the temperature, the more the liquid formed, and the more the material shrinks. In practice, the heating must be restricted lest so much liquid forms that the whole brick starts to become distorted under the weight of the higher layers of bricks. In extreme cases, the bricks get fused together in the kiln. Gas formation can 'bloat' brick faces.

A few hours 'soaking' at the finishing temperature is recommended to ensure that the whole brick has attained uniformity. New materials, such as mullite, may crystallise from the liquid at temperatures which may reach approximately $1,100^0C$ for some brickmaking clays. In these ceramic reactions, a long firing time at a low temperature can have the same effect as a shorter firing-time at a high temperature. As cooling commences, the liquid solidifies to glass, bonding other particles together. The cooling rate should be slow to avoid excessive thermal stresses in the bricks, particularly once the quartz inversion temperature (573^0C) is reached, since shrinkage occurs in the presence of quartz.

The inevitable firing shrinkage should be fairly small, otherwise it would be difficult to maintain the stability of the bricks in the kiln.

II.6 Other basic requirements

High technology tends to limit the range of clay types acceptable for a particular process machine, and is less versatile as regards the type and grade of fuel. On the other hand, a wide range of materials and fuels can be used with less sophisticated technologies. Fuel, whether oil, gas, coal, wood, scrub or plant wastes, must be available for the brickmaking process and may be regarded as a raw material. Electricity may be advantageous for ancillary purposes. Water is also necessary and, for highly plastic clays, sand may also be required.

III. CLAY TESTING AND SIGNIFICANCE OF RESULTS

Although highly sophisticated clay testing methods have been evolved, very simple tests can also give useful information. The former may be necessary for large turnkey projects, where equipment is often adjusted for specific raw materials characteristics. However, they require skilled staff not only to carry out tests, but also to interpret the results. On the other hand, simple tests may often be carried out on site, by less qualified personnel. Yet, the results may be more easily related to the use of the raw material than those obtained from more sophisticated tests.

The most direct test method used successfully for thousands of years involves visual inspection and the feel of the soil, and the carrying out of brickmaking trials.

Tests to investigate various aspects of a soil's suitability for brickmaking are given below, starting with the most basic field test methods. Simple, intermediate technology tests are described next. Finally, a brief description of the more sophisticated tests which might be employed if adequate facilities exist, is provided at the end of this section.

III.1 _Particle size_

A visual inspection of the raw material will show whether the soil contains sand; a magnifying glass may assist in this operation.

The 'feel' of a soil in the hand will give an indication of the proportion of different particles sizes. When dry, a sand constituent gives a sharp gritty feel. A piece of the hard soil rubbed with the back of the finger nail cannot be polished. When wetted and broken down between the fingers, the sand particles become more readily visible.

If there is a high proportion of clay the dry soil will feel smooth and powder may be scratched off it. Furthermore, a surface of a small lump can be polished with the back of the finger nail. Damp soil can be worked into any shape, but will tend to stick to the fingers. The more clay in the soil, the more difficult it will be to remove it from the hands by wiping or washing.

A suitable brickmaking soil will have a high proportion of sand, so that it may not take a polish. High clay content soils may need addition of sand to make them suitable.

An estimate of the proportions of the various size fractions can be obtained using the sedimentation jar test. Any straight-sided, flat-bottomed, clear jar or bottle may be used. An approximately one litre capacity jar will be adequate (figure II.4). One-third of the jar is first filled with broken-up soil. Clean drinkable-quality water is then added until the jar is nearly full. The content of the jar is next mixed up, one hand covering its mouth to avoid spilling. The soil is then left to settle for an hour, shaken again and allowed to settle a second time. An hour later, the depth of the separate layers can be seen and measured. The bottom layer consists of sand and any coarser particles. The medium layer consists of silt and the top layer of clay. Often, the top two layers will merge together. The settlement of the clay fraction may be slow with some soils. The use of salty water for this test will flocculate the clay and help it to settle, thus giving a clearly defined level in the bottle which can be measured more easily.

Where laboratory facilities exist, a wet sieving process may be used to estimate the quantities of various sizes of sand. The soil is first washed through a nest of sieves of increasingly fine mesh, and the quantities retained on each sieve are dried and weighed. The difference between the weight of these fractions and that of the initial sample is then equal to the weight of silt and clay. Further information about the composition of these finer materials can be obtained using a sedimentation method (the Andreason pipette) or a hydrometer method. Details of these and other methods are described in British Standard Methods of Test for Soils for Engineering Purposes - BS 1377:1975 (18).

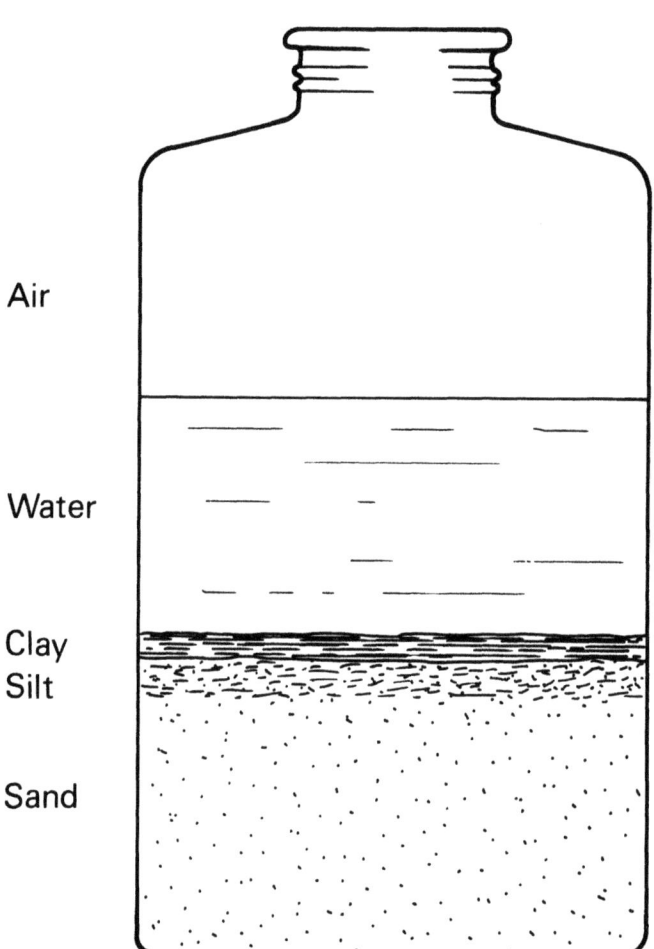

Figure II.4

Jar test

Soil used for the production of bricks by traditional methods should contain the following:
- 25 per cent to 50 per cent of clay and silt; and
- 75 per cent to 50 per cent of sand and coarser material.

The soil should preferably contain particles of all sizes

III.2 **Plasticity and cohesion**

If the moistened soil is rolled by hand (on a flat surface) into a cylinder, a sharp break of the latter when pulled apart indicates a very sandy soil with low plasticity(7). On the other hand, the soil may be considered adequate for brickmaking if the cylinder elongates to the point of forming a neck before breaking

Another test consists of preparing a long cylinder of 10 mm diameter and letting it hang unsupported while holding it from one end. The length of cylinder which breaks off will provide fairly accurate information on the properties of the soil. The breaking-off of a piece of cylinder of 50 mm or less will indicate that the soil is too sandy for brickmaking. In this case, it will be necessary to add some fat clay or ant hill material to the soil. On the other hand, the breaking-off of a piece of 150 mm or more will indicate the presence of too high a proportion of clay, necessitating the addition of sand or grog to the soil. A soil adequate for brickmaking will require that the length of the broken-off piece of cylinder is between 50 mm and 150 mm (19).

The properties of the wetted soil will depend upon the moisture content. A ball of suitable soil containing the correct amount of water should break into a few pieces when dropped from the held-out arm on to hard ground. On the other hand, a flattening out of the ball will indicate that the soil is too wet, while the breaking of the ball into a large number of small pieces will indicate that the soil is too dry. Some more precise assessment of plastic properties can be obtained by simple laboratory tests. The soil should be mixed up with an excess of water to make a very runny paste or slip. The latter is then poured on a dry porous plastic plate, and mixed continuously with a spatula or knife. As water is absorbed by the plate, the soil will become less liquid and new incisions made with the knife will take longer to close. Once an incision remains open, approximately 5 g of material should be taken from its vicinity and weighed immediately. The sample is then weighed again after a few hours' drying in an oven at $110^0 C$. The moisture content can thus be calculated as percentage of the dry weight of clay. This percentage is termed the liquid limit of the soil.

Some small pieces of the clay may be removed from the plate and rolled by hand of a flat-glass plate in order to make filaments approximately 3 mm in diameter (figure II.5). At first, long filaments may be fashioned easily. Then, as the soil dries out there will come a point when they just start to crack longitudinally and break up into pieces approximately 10 mm long. Once this occurs, approximately 5 g of such pieces should be weighed, oven dried, and weighed again to determine the moisture content as a percentage of the dry weight of clay. This percentage is termed the plastic limit of the soil.

The difference between plastic and liquid limits is the plasticity index.

When more advanced facilities are available the liquid limit should be determined with the cone penetrometer, described for example in BS1377 (see section III.1). In this test, the penetration of the point of an 80 g metal cone having an apex of $30°$ is measured as it rests for 5 seconds on the moistened soil. From a series of readings for different moisture contents the liquid limit is determined as the moisture content which gives a 20 mm penetration. The test for estimating the plastic limit is the same as that described above.

Several other testing methods are used in well-equipped laboratories (17).

Soils with a low plasticity index will be difficult to handle for brick-moulding: the green brick will distort after demoulding if the soil contains a small excess of water while the soil will be too stiff to mould if it lacks sufficient water. A high plasticity index is therefore preferred.

Soils with a high plastic limit will require a great deal of water before they can be ready for moulding. Long drying is then necessary prior to firing. A high plastic limit and very high liquid limit may indicate the presence of montmorillonite, with its attendant moisture movement problems. Thus, montmorillonitic soils may not be adequate for simple brick-moulding methods as the latter require a relatively high moisture content. They need either high compaction pressures on semi-dry mixes, or dilution with non-shrinking materials. Montmorillonites may give rise to size changes in the drying bricks as the humidity of the air varies naturally.

In a recently published book (20) reference has been made to an earlier suggestion (21) that, within the plasticity ranges indicated in table II.2, a soil may be adequate for the production of bricks by traditional methods. However, it may be possible to use materials with plasticity limits outside the ranges shown in the table.

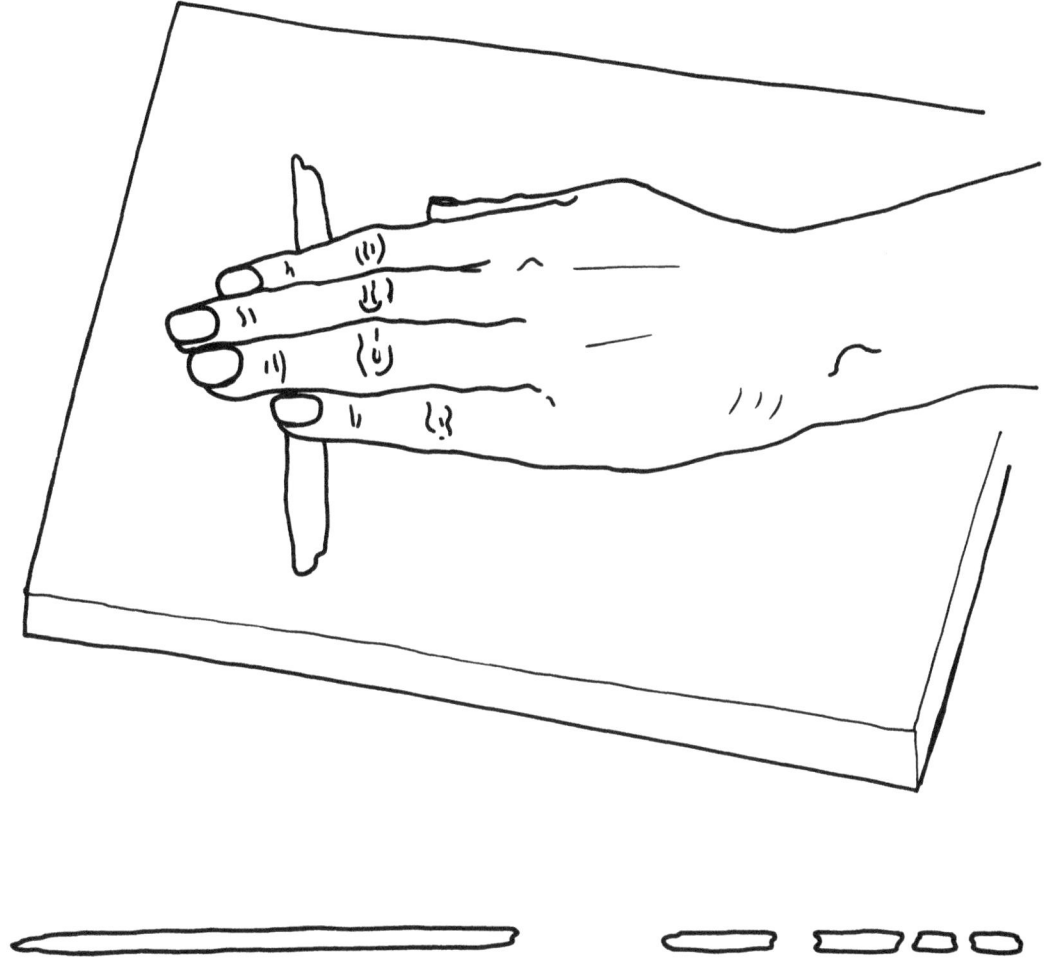

Figure II.5

Plastic limit test

Table II.2

Plasticity limits for good brickmaking soils

Plastic limit	12 to 22
Liquid limit	30 to 35
Plasticity index	7 to 18

III.3 Mineralogy and geology

The mineralogist recognises the presence of certain minerals in the field while the geologist identifies structures in the earth's appearance that will assist in locating suitable raw materials sources.

The work of the mineralogist will consist largely of taking samples from the field and examining them under the microscope in a laboratory. On the basis of information from other tests, he may identify the components of a soil and thus determine their suitability for brickmaking by the various means available. In more advanced laboratories, the electron microscope (especially the scanning electron microscope) will be a useful tool. Identification of minerals will also be greatly assisted by X-ray diffraction analysis.

III.4 Chemical analysis

The colour of samples of materials obtained from field investigations gives some indication of the composition of the soil. Red soils may be high in iron, which can act as a flux. Very dark colours, or a musty smell in the damp soil, may indicate the presence of organic matter: it may be possible to use such soils, though their agricultural use should be given first priority. Dried out encrustations on the surface of the ground indicate the presence of soluble salts, which are best avoided for reasons given in Section II.3.

A simple laboratory test for the presence of sulphates consists of dissolving these salts and adding a solution of barium chloride. The forming of a white precipitate will indicate the presence of sulphates. On the other hand, chlorides can be detected by addition of silver nitrates. These chemical tests could be done on site, with a small portable test kit. The presence of calcium carbonate can be ascertained by the existence of lumps or nodules which are likely to be white, or by effervescence from gas produced by the addition of dilute hydrochloric acid to the soil. An estimate of the quantity of carbonate has been suggested(7): 1 to 8 per cent in case of slight bubbling; 8 to 16 per cent in case of pronounced bubbling; and 18 per cent in case of sudden foaming.

In a properly-equipped laboratory, a full chemical analysis may be undertaken, which, together with the mineralogical examination, can assist in identifying the constituents as mentioned in section II.3.

III.5 Drying shrinkage

High clay content (recognisable in wet conditions by the stickiness of the soil) is in dry weather, recognisable by the presence of shrinkage cracks in exposed soil, in either vertical or horizontal faces (see figures II.6 and II.7).

To obtain a measure of the shrinkage of a moist soil, which may seem suitable for brickmaking, the most simple method is to mould a few bricks from the soil and allow them to dry thoroughly. The length of the dried bricks and of the moulds are then measured in order to obtain an estimate of the linear drying shrinkage. The latter may be obtained from the following formula:

$$\text{Linear drying shrinkage (per cent)} = \frac{(\text{Mould length} - \text{final dry length}) \times 100}{\text{mould length}}$$

The appearance of the test bricks will give some indication of the suitability of the soil for brickmaking. It is suitable if no cracks appear on the surface. If some slight cracks appear it would be advisable to shorten the soil by adding 20 per cent sand or grog. In case of extensive cracks, 30 per cent might be mixed in. Soil too lean for moulding will have to be made more fat with other clays, or ant hill soil.

Generally, up to 7 per cent linear shrinkage may be tolerable, depending upon the nature of the material and the rate of drying. If linear shrinkage is more than 7 per cent shortening is advisable(22). In any case, it is necessary to know the linear shrinkage in order to determine the exact size of moulds needed for producing bricks of given dimensions.

If more organised test facilities are available, it would be advisable to prepare special shrinkage bars. For this test, an open-topped wooden mould, approximately 300 mm long by 50 mm deep and wide, should be made up by a carpenter or a sufficiently skilled handyman (figure II.8). The soil used in the test should be dried, if not already so, and broken down. Large stones should be removed. It is then mixed with just sufficient water to bring it near the liquid limit (i.e. pieces of the soil should be deformable yet retain their shape). If time permits the soil should be covered, left overnight, then mixed up again. The mould should be lightly greased inside to prevent the soil from sticking. Some moist soil is then laid in the bottom, and the mould tapped on the bench or ground to cause entrapped air bubbles to escape from the soil. The mould should be filled in the way described above in several stages, and excess soil struck off the top to leave a surface level with the surface of the mould. The soil should be dried slowly at first, at room temperature. Once shrinkage appears to have stopped, it may be tipped out

Figure II.6

Clay shrinkage on a vertical face

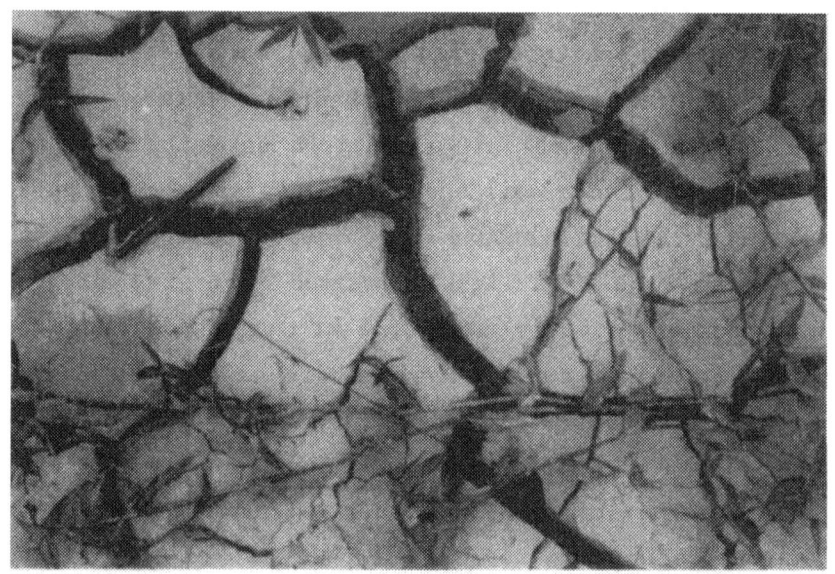

Figure II.7

Shrinkage cracks in clay pit bottom

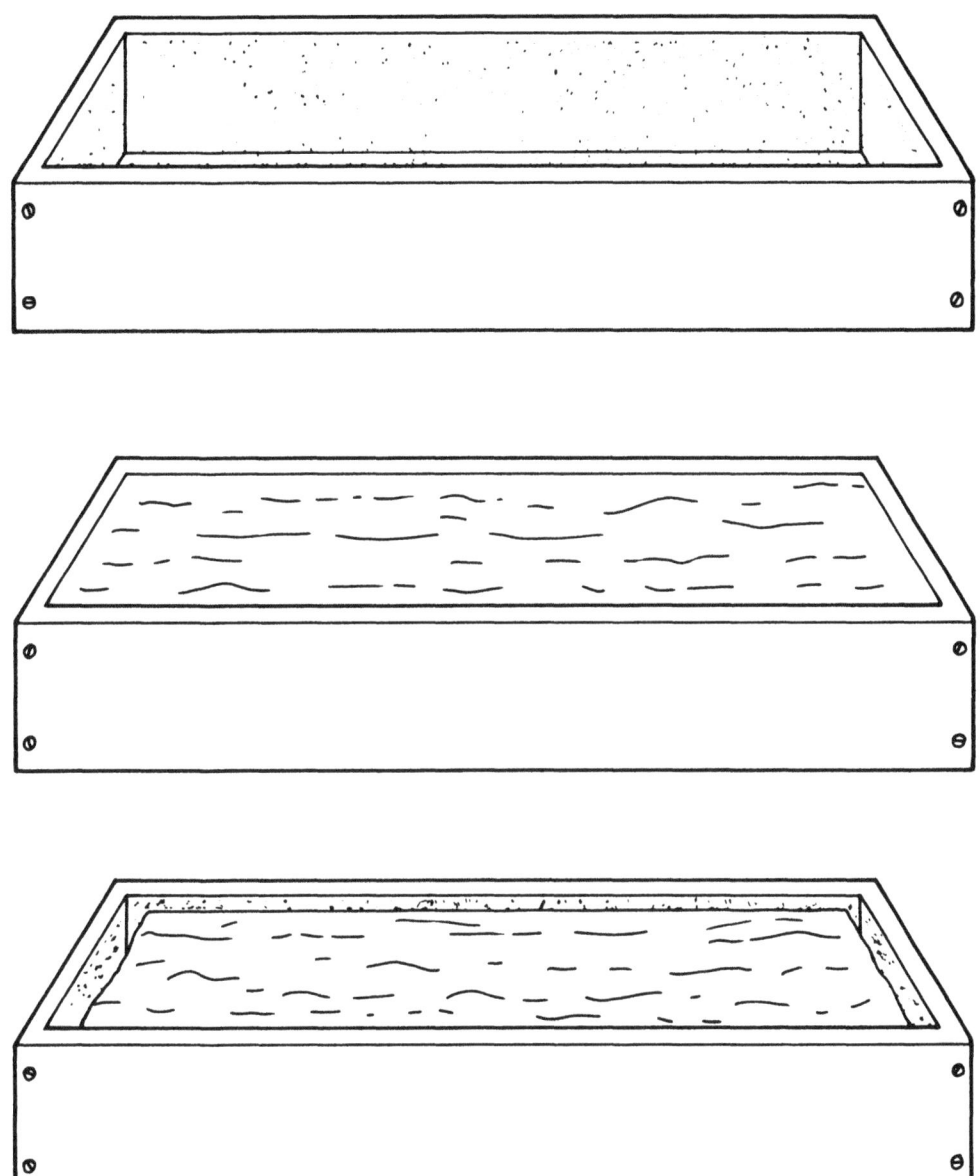

Figure II.8

Shrinkage mould

the mould and dried in an oven at 110^0C. The linear shrinkage may then be calculated as indicated above.

III.6 Firing shrinkage

Some shrinkage during firing is inevitable. From 6 to 8 per cent linear shrinkage is desirable (5,7). The simplest field test to measure firing shrinkage is to burn a whole batch of bricks.

Measurements of firing shrinkage are more readily obtained in the laboratory than in the field. Small bars should be moulded, dried, measured, then fired to various temperatures in a laboratory furnace. They are then cooled and re-measured to calculate the linear firing shrinkage. A special furnace has been designed for this test. It requires only one sample, since it has a horizontal silica rod whose movement is measured outside the furnace as the temperature rises. A 'gradient' furnace of uneven temperature distribution can also give useful information.

CHAPTER III

QUARRYING TECHNIQUES

I. <u>ORGANISATION AND MANAGEMENT OF THE QUARRY</u>

The quarry should be located in an area with sufficient proven deposits of good brickmaking soil and, preferably, a thin layer of overburden to minimise excavation work. The operation of mining clay from the clay pit or quarry is generally referred to as 'winning the clay'.

I.1 <u>Opening up the quarry</u>

Access to the quarry from the bricks production plant should be quick and easy, preferably with no more than a slight gradient. A good route will minimise effort, time and expense in transporting clay, and will facilitate supervision of the pit. A track or roadway may need to be constructed, especially if wheeled vehicles are to be used to convey the clay to the brickworks. Trees and bushes must be cleared, and may be sold or kept for fuel.

Prospecting will indicate whether the clay stratum is horizontal or sloping. If it does slope down into the ground, the worker should face that direction and remove the top soil. The top soil should be piled in two rows along the excavation. The trench thus formed will have a horizontal bottom, along the 'strike' of the clay stratum(24). The angle at which the clay stratum slopes from the horizontal is a measure of the "dip" of the stratum. It must be borne in mind that more overburden will have to be removed as the clay winning proceeds. If there is no dip, the trench may be dug in any direction.

As a general principle it is unwise to start digging for clay at the lowest part of the ground(25) since surface water from rainstorms will then immediately flood the clay pit and stop the work. It is preferable to start digging at a higher point. This should be borne in mind whether the underlying clay has a dip or not.

A sufficient area of overburden should be removed to prevent any of it falling into the clay as winning proceeds (e.g. up to 10 m may be taken off to each side of the trench). If too much is cleared weeds may start to grow and will have to be cleared again. Clay may then be dug to a depth of a few metres, along the centre line of the exposed area. The actual depth will depend upon the adopted method of digging and the nature of the material extracted. Further material is then obtained by widening this deeper trench a small amount at a time. Eventually, it will be necessary to remove more overburden. This unwanted material may conveniently be used to fill the first-opened part of the trench once all the useful clay has been excavated. If good clay does extend lower down, it might be extracted at a much later date or by the method of 'benching' or 'terracing' (which is the working of two or more clay faces at different depths at the same time).

I.2 Operating the quarry

Safety

It is important to bear safety in mind in the clay pit. Clay is very slippery when wet. Thus steeply sloping paths and access routes should be avoided. Steep drops into the pit would be hazardous. Damp clay is not stable at a near vertical face. Consequently, a whole portion of the material may undergo a rotational slip, into the bottom of the pit. For this reason, it is advisable to slope back or 'batter' the faces of the pit as they are dug. The latter should not be too high.

Water

If flooding of the pit bottom becomes a problem, the water may be drained away through a downhill channel. If this is not possible, a sump must be dug in the pit bottom to collect the excess water. The latter may then be removed from the sump by pumping or with buckets. This water may be contaminated with soluble salts present in the ground. Thus, it would be unwise to use it in the subsequent brickmaking process, unless tests show that it does not contain salts. In countries having a wet season or monsoon, the quarry may need to be abandoned until the rainwater is drained off.

Rejection of impurities and reinstatement

As digging proceeds, the workers should discard any plant roots, stones, limestone nodules or harder clay inclusions since they would cause problems in subsequent processing. Any pockets of unsuitable soil should be removed rather than left in place and the pit should be kept tidy. The top soil should be returned ultimately, on top of discarded material into the worked-out part of the quarry. Crops may thus be grown again. This is the case, for example, in Madagascar where the most exploited clays for brickmaking are from the rice fields. The top soil is then reinstated for rice growing (26). One of the main sources of raw materials for the structural clay industry in Indonesia is also the rice fields.

Rate of extraction

The rate of extraction of clay from the pit must be sufficient to meet the demands of the brick-moulders. Alternatively, it may be slightly larger in order to guard against problems which may arise unexpectedly in the pit, such as temporary flooding, presence of an unsuitable pocket of material, contamination of the clay, etc. In some countries the onset of the wet season or the monsoon may halt operations in the clay pit. During these times, the natural drying of moulded bricks will become almost impossible, building of field kilns or clamps (Chapter VII) will be impracticable, and demand for bricks will fall due to adverse weather conditions restricting building and construction activities. In such cases, the pit will be closed and the whole brickmaking operation stopped. However, in other places, although the rain may prevent operations in the clay pit, some demand for bricks may continue, and it may be possible to carry on brickmaking and drying under cover. More permanent forms of kiln, also under cover, may still be in operation. In these cases, sufficient clay should be won from the pit during the dry season, and stock-piled, to meet the demand when no more can be mined. In some communities the workforce may wish to engage in agricultural activities during the harvesting season. This factor should be taken into account in designing the whole brickmaking process.

Working of the clay face

The depth of the top-soil may vary from a few centimetres in some arid climates to several metres in hot, humid areas. Frequently, a layer of sand may occur below the top-soil and over the clay layer. The best clay for brickmaking is likely to be that immediately below the sand, since it is likely to contain a proportion of sand itself. However, the depth of this good material may be small. Clay lower down may be too fat and will need addition of sand from above. Hence the best method of operation is to work a quarry face in such a way as to dig both clay and sand, taking shallow slices down the face, to obtain a suitable mixture (see Figure II.6). Another virtue of taking shallow slices of the face is that any embedded stones can be found more easily than if large cuts are taken. These stones can then be discarded(28).

If suitable soil containing the desirable proportions of sand and clay cannot be dug at one face only, it may be necessary to obtain a fat clay from one face and sand or sandy clay from another. This has often been done as, for example, near Mombasa in Kenya, where material from two faces has been mixed in the pit bottom prior to use. If the material varies horizontally (i.e. from one place to another) two separate faces in the pit, or two separate pits, might be worked simultaneously. If the material varies vertically (i.e. at different depths), two faces can be operated by benching (see section II.2).

Record-keeping

For later reference, a note-book should be kept for recording progress, and any significant happenings in the quarry. A map should be made of the quarry, showing the position of original test holes or pits, the depth of clay, and other major features such as streams, tracks, large trees and the brickworks if adjacent. The position of the clay-pit face should be drawn on the map every few months, and the date written on the line representing the face. If the floor of the pit is dug a second time or if benching is used, a second colour could be used to update the map. This will assist in an orderly exploitation of the reserve: haphazard digging is wasteful of material and effort[27]. The rate of ingress into the reserve should be clearly visible, and if problems or complaints arise with the finished bricks, the fault may be traceable to a cause in the pit. The extent of any problem materials in the pit should be marked on the map. The supervisor should check constantly the work at the clay face and inspect the material being won to ascertain that it is suitable and does not contain deleterious materials.

II. METHODS OF WINNING THE CLAY

Two basic methods are available : mechanical winning and hand-digging. These are briefly described below.

Mechanical winning

Mechanical methods such as the use of the drag-line and multi-bucket excavator are mostly appropriate for the largest-scale brickmaking operations. It is most unlikely that even a face shovel (figure III.1) could be justified in works of the size considered in this memorandum, unless it is available on hire from a nearby depot for a short period of time each year, (e.g. in order to build a stock pile). It seems unlikely that mechanical winning could be economical for output of less than 14,000 bricks per day[22].

On the other hand, the more commonly available and versatile bulldozer could have a place in the laborious task of clearing overburden on infrequent occasions. It might be brought in on hire, or when available from nearby road construction or other civil engineering works (e.g. against payment of a fee).

Most of the clay resources utilised by the small-scale manufacturer are likely to be of the soft plastic type. In some areas, when only hard shales are available, blasting might be undertaken occasionally to loosen material from the quarry face.

II.2 Hand-digging

Hand-digging has been widely used even for medium-size production plants, because of its versatility in dealing with all clays from soft muds to shales or even with ant hills. Hand-digging can also be adjusted to various types of work, and allows workers to sort out unwanted stones, limestones, roots, etc. It also avoids large amounts of capital investment, the stocking of spares and the organisation of maintenance of machinery. In many situations, hand-digging may be the only possible means of winning clay.

The rate of winning clay will depend upon the type of clay, the nature of the pit and the productivity of labour. Productivity rates for one man digging enough clay for the production of approximately 3,500, 1,500 and 4,000 bricks per day have been estimated(5,25,8). However, these estimates are not strictly comparable as some of them include an element for the transport of clay over a short distance. Measurement of shovelling rates in the American mines(29) indicated an optimum working day of 6.5 hours. Longer working hours result in lower outputs.

Once clay has been dug, there will be a natural reluctance to reject any which may prove unsuitable, especially after the hard work of winning it. In particular, the workers paid according to quantity excavated may be reluctant to reject unsuitable material. Hence the importance of supervision, inspection and quality control.

If the face is benched, the separate levels need be only 1 m different(25) and 0.5 m wide(5), especially if materials from two or more levels are to be mixed. This can be done by throwing all materials down to the lowest level for mixing, and subsequent transportation away to the works.

The details of working the pit must be decided locally. For example, at Asokwa in Ghana (figure III.2) the clay was hand-dug from a face which was approximately 2.5 m high in places(30).

Steel bladed, medium-weight spades are well suited for digging plastic clays. Preferences in blade design vary from country to country. It is,

Figure III.1

Face shovel or single bucket excavator

Figure III.2

Clay-pit at small brickworks (Ghana)

however, recommended to use narrow and slightly conical blades for the digging of this type of clay. The handle of the shovel should be shorter whenever the foot is used on the top of the blade. In places where this is not done, the handle is traditionally very long.

If hard, dry clays are to be won (figure III.3) it may be necessary to loosen them from the face with a pick, then shovel the material away.

In many countries, the hoe and mattock are more generally used and are suitable for winning clay.

III. <u>TRANSPORTATION TO THE WORKS</u>

In large works, clay is conveyed from the pit in various ways, including the use of lorry or truck, large dumper truck, small-gauge railway systems, aerial ropeway or belt conveyor. Capacity, capital cost, maintenance and repair militate against the use of these methods for the smaller works.

While heavy transport equipment may not be suitable, the use of a small diesel-powered dumper or a front-end loader may become feasible on a hire basis. Similarly, an agricultural tractor may be used to haul a loaded trailer of clay. Alternatively, a draught animal may be useful for pulling a trailer if the road is not too steep and muddy.

The wheelbarrow is a versatile and low-cost device for moving clay. It need not be all-steel or specially imported and can be locally produced from available materials. It can be taken from the clay face to any desired point at the plant site, on a narrow path or a plank on muddy ground (figure III.3). The larger the wheel, the more easily the barrow will pass over irregularities in the ground. The wheel should be as close to the load as possible in order to take the weight off the hands. The handles should also be approximately 50 mm lower than a standing person's palms when the barrow is at rest on the ground. Thus, the arms are just slightly crooked when the barrow is wheeled.

A simple aid for carrying clay and other materials is the litter (figure III.4). Its use by two people avoids twisting the body. Large loads may be carried over rough terrain, or up steep slopes. It may be fabricated easily by unskilled labour using cheap, locally available materials.

The most simple devices for transporting clay are the basket and headpan, both of which will be available in many communities.

Figure III.3

Digging and transporting of dry hard clay (United Kingdom)

Figure III.4

Litter for carrying wet clay (Sudan)

CHAPTER IV

CLAY PREPARATION

Good bricks of consistent quality and free from defects will only be obtained if the materials used are of suitable and uniform nature. Only rarely will such a material be won directly from the clay pit. Commonly, some preparation and pre-processing is necessary to remove unwanted inclusions, add non-clay materials, or mix the materials for uniformity. In most cases, water must be added in order to bring the clay to a suitable consistency for forming into shape.

Adequate careful preparation will mitigate the following problems which might otherwise arise with the bricks.

- Overall cracking on the surface due to the use of too high a proportion of clay fraction in the mix. This may be minimised by adding sand or grog.
- Localised cracking over a hard piece of clay (figure IV.1), or a large stone; mitigated by crushing or removing such inclusions.
- Limeblowing (figure IV.2) which may be avoided by removing the larger limestone pieces and reducing any remaining limestone to less than 2 mm across pieces, or smaller pieces if the quantity of limestone is high. Indian research(31) suggests that the addition of 15 kg common salt added to the clay per 1,000 bricks should minimise limeblowing.
- Low green strength of dried brick, possibly due to insufficient clay in the mix.
- Lack of plasticity, making the forming process difficult; also due to insufficient clay in the mix.
- Non-uniformity of size, shape and strength due to insufficient mixing of the materials.
- Efflorescence and sulphate attack of cement-based mortars (see Chapter VIII), which might be reduced by rejecting surface clay where the salts may have accumulated naturally. There is also an expensive option of chemical treatment with barium carbonate. It is extremely difficult to wash salts out of clay, as it is difficult to mix the latter with water. Furthermore, large quantities of water will be required for this purpose.

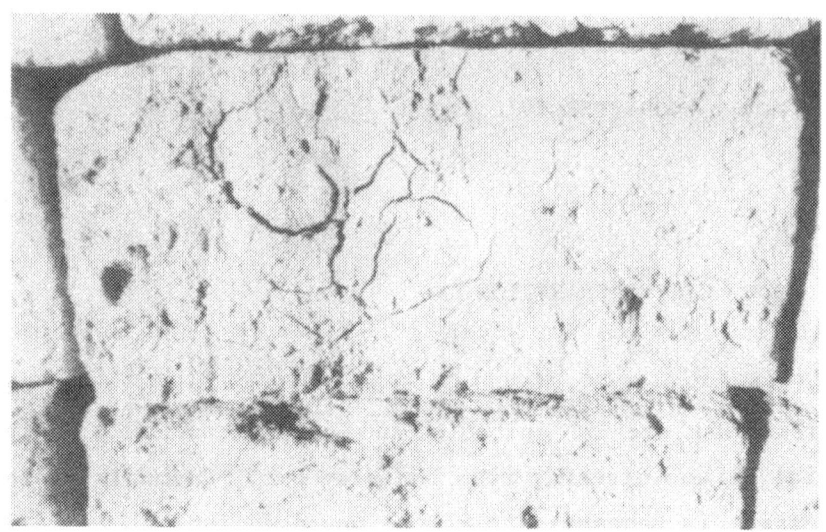

Figure IV.1

Localised cracking due to inadequate clay preparation (East Africa)

Figure IV.2

Limeblowing in bricks (Central Africa)

Efflorescence on bricks made from clay containing high concentrations of soluble salts is shown in figure IV.3 In some circumstances, salts crystallise beneath the surface, buidling up stresses which can force flakes to spall from the surface as shown in brickwork in a boundary wall in India (figure IV.4).

I. **MAIN CLAY PREPARATION PHASES**

Clay preparation includes the following operations:
- sorting (or picking) and washing;
- crushing or grinding;
- sieving or screening;
- proportioning;
- mixing, wetting and tempering.

A whole range of motor-driven machines is available for these operations, including belt-conveyors, jaw-crushers, kibblers, hammer mills, grinding pans (both wet and dry), rolls, de-stoning machines, vibrating wire screens, proportioning feeders, double-shafted trough mixers, and pug mills. However, few of these capital-intensive items will be appropriate to the type of production described in this memorandum. In a labour-intensive set-up, a mixing machine may be the most useful piece of equipment if diesel or electric power sources can be used. Animal power may also be worth considering.

It is best to prepare clay in a very dry or a very wet condition. Damp clays are difficult to crush, they stick on sieves, are awkward to handle and require much power to mix.

II. **SORTING**

An essential part of clay preparation is that carried out in the pit. This includes the discarding of unsuitable pockets of soil, roots, stones, limestone nodules, etc. and the winning and preliminary mixing of clayey and sandy materials. Visual inspection of the clay in the works is not easy to carry out or enforce, but is done on a routine basis whenever the clay can be moved on a narrow conveyor belt past workers who pick off any unwanted material. It is advisable to have the supervisor check the clay coming into the works from time to time. Unwanted materials detected at this or any subsequent stage should be removed.

Where stones or limestone nodules constitute a particular problem they can be removed in a washmill (see Section IV.3).

Figure IV.3

White efflorescence on brickwork (Middle East)

Figure IV.4

Flaking of the type produced by soluble salts (India)

III. Crushing

In the tropics, clay will generally be dry when won from the pit. Thus the centres of large lumps will be difficult to wet. Non-uniform material is likely unless the dry clay is first crushed to less than a few millimetres across. Where capital cost is justified by a sufficiently large production scale, and where power sources are reliable, crushing rolls may be useful. Figure IV.5 shows crushing rolls in Ghana.

III.1 Manual pounding and the hammer hoe

Manual pounding with a hammer or punner may be used in small works but is very laborious. There is a tendency for already broken pieces to be compacted again, forming a hard lump which prevents the tool from breaking fresh material. It is thus necessary to clear away material as soon as it is broken. In favourable circumstances, two tonnes might be prepared per day by a team of four men (e.g. enough for 1,000 bricks).

The hammer-hoe (figure IV.6), which is used in Malawi, is a useful dual-purpose tool, having special uses not only in the works, but also in the clay pit. Material can be won, turned over, and mixed with the hoe. If hard lumps are found in the mix, it is not necessary to exchange tools as a half-turn rotation of the handle will bring the hammer into position for breaking the lumps.

III.2 The pendulum crusher

A labour-intensive crushing machine has been developed by the Intermediate Technology Workshop in the United Kingdom especially to meet the needs of the small-scale brick-maker as identified in an earlier survey(10). It is easily built from mild-steel sections, and works on the pendulum principle. The soil, which is placed in a feed hopper at the top of the pendulum, comes into contact with a static grinding head and a curved moving grinding head. The latter is attached to the top of the heavy pendulum which is kept swinging by two people (figure IV.7). The moving head is studded with protruding bolt heads which entrap and crush clay as the head rotates in a downwards direction. Ground clay falls through by gravity on to a built-in sieve which can be of any desired mesh size. On the upward return move, any remaining clay is cleared from the grinding surfaces prior to the next downward swing, so that a slight dampness of the clay is not a great problem. Figure IV.8 shows details of the components of the crusher.

To operate the machine, two men start the pendulum swinging. Once the latter has reached a maximum angle, a third man starts feeding material. If exceedingly hard pieces are encountered or if, for example, a steel tool is

Figure IV.5

Motor-driven crushing rolls (Ghana)

Figure IV.6

Hammer hoe

Figure IV.7

Manually-powered pendulum crusher

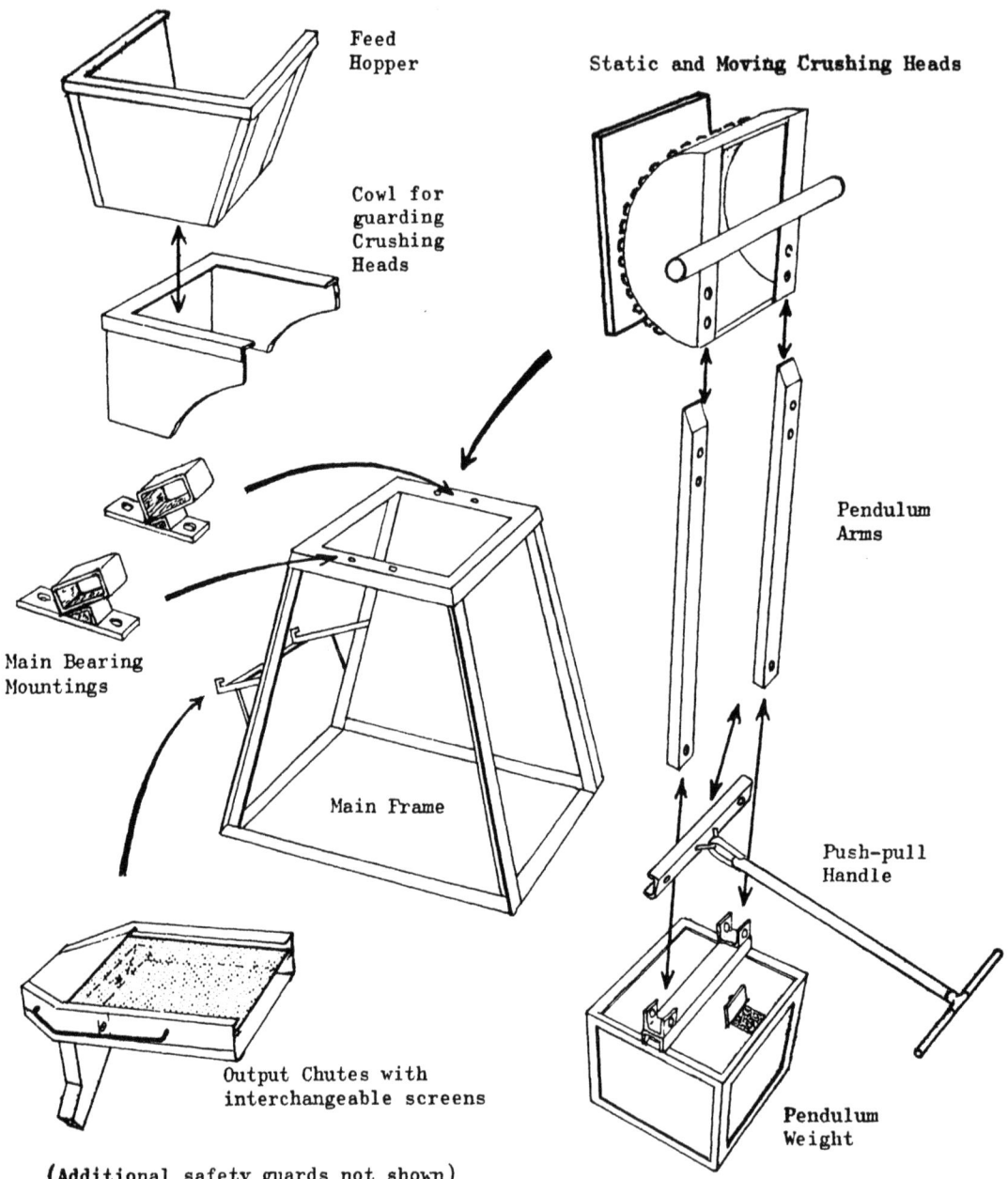

Figure IV.8

Manually-powered pendulum clay crusher-
Detailed drawing of parts

accidentally dropped in the hopper, the grinding heads will not be damaged, nor will the machine stop since the pivot bar of the pendulum can run up in the rectangular bearing box, as a safety measure. The whole machine is guarded to minimise the risks to operators. A fourth man should be available to remove the ground clay and return rejects from sieving to the pile of clay yet to be crushed.

Periodically (e.g. each time the container is full of finely ground clay) the operatives would be well advised to change tasks on a regular rotation, moving between jobs as follows: 1: feeding clay; 2: pendulum handle (right side); 3: pendulum handle (left side); 4: attending discharge and resting, then back to feeding again.

An extensive series of tests has shown some variation of production rates depending upon clay type. On average, a four man team may produce crushed clay at a rate of 20 tonnes per day, (i.e. enough for 10,000 bricks). This was an average for the easier alluvial and sedimentary clays. If harder shales are used, enough material may be produced for 8,000 bricks per day. The same team may also prepare a tonne of grog by crushing underfired bricks in 2.25 hours.

Occasional greasing of the bearing of the pendulum is the only servicing necessary. From time to time the machine should be inspected for wear. The pivots and the bolts protruding from the moveable head are the parts most likely to deteriorate. They should be simple to replace when worn out.

The pendulum crushers, which have been operated in several countries, can be fabricated from readily available steel sections. None the less, entrepreneurs should first refer to the innovators for precise details before adopting the method. Ready-made machines or sub-assemblies can also be obtained from them (see Appendix IV). Manufacture under licence is being arranged in several countries.

III.3 Animal-powered roller

A traditional crushing method in India uses a heavy stone roller pulled by a bullock over the dry clay. The latter is laid out in a circle at the centre of which the axle for the roller is pivoted. Other draught animals could perform this task.

IV. SIEVING

Crushed clay must be sieved to ensure that over-size pieces are not used. These should be returned for further crushing. The finer the material the better the quality of the brick. However, clay passing through a 5 mm sieve should be satisfactory. The simplest device for sieving is a screen of wire mesh, fastened on to a rectangular frame resting on the ground at one end and supported on legs at the other at a 45^0 angle to the ground (figure IV.9).

Figure IV.9   Sieving

V. PROPORTIONING

The mixing of two different dry materials is best undertaken after crushing and sieving. Quantities of clays, grog, sand, etc. should be measured by volume since this method is generally much easier than weighing. Measurement is advisable in order to get a consistent product quality. Measures should be taken from each constituent in turn, to give a preliminary amount of mixing.

VI. MIXING, WETTING AND TEMPERING

Dry ingredients may be mixed with hoes or spades. In addition to clay, sand and grog, solid fuel may be mixed with the clay to assist with the burning of the bricks (See Chapter VII). Mixing should be continued until no difference of colour or texture can be seen from place to place in the pile. For special purposes, where soluble sulphate salts constitute a problem, barium carbonate (precipitated) may be added. A fixed quantity, such as 10 kg per 1,000 bricks, may not be a priori recommended since it will depend on the extent of the problem.

Water must be added to most soils, over the whole surface of the spread-out material, or as each layer is deposited in a pile. Water should be sprinkled with a watering can fitted with a rose spray, or a similar device.

Mechanical power may be used for the laborious task of mixing and kneading the clay. Concrete mixers have a rotating drum with paddles attached inside. The consistency of wet concrete is such that the paddles can just about pass through, and any adhering concrete slips off when the paddles rise. Consequently, only an extremely wet clay, too wet for brickmaking, could be mixed in a concrete mixer as stiffer mixes will merely adhere to the paddles and drum. For stiff mixes it is best to use a stationary container equipped with rotating paddles or blades. This should be borne in mind by any person attempting to design clay-mixing equipment. For example, an oil-drum rotating on bearings fixed to each end will generally fail to achieve effective mixing.

A motor-powered double-shafted trough mixer is ideal, and readily available from major equipment suppliers. However, acquisition of this equipment involves high capital investments. The capacity of this equipment may also be too large for small production units. Single-shaft mixers are not so efficient. Trough mixers need to be very strong and powerful, and may be unsatisfactory if made of too thin a gauge steel and undersized bearings. They are made and used in Ghana by a number of brickworks (32).

Much of the work of mixing water and soil may be avoided by the simple expedient of waiting for the water to percolate right into the structure of the clay. The thoroughly mixed dry constituents, having been wetted and piled up as described above, should be covered with sheeting or sand to prevent evaporation and drying. This long-term wetting process, known as "tempering" or "souring", allows chemical and physical changes to take place in the clay, thus improving its moulding characteristics. These benefits may be achieved within a day or two with many clays, though others (such as the harder shales) may require weeks.

VI.1 Pugmills

A pugmill is a most useful machine for mixing wet clay ready for moulding. High capital-cost machines from the major suppliers of brickmaking equipment have a series of angled blades rotating on a horizontal shaft within a closed barrel. These blades mix the clay and force it out by an opening in the barrel. Electric or diesel power are used for this type of machine (figure IV.10).

Cheap, animal-powered pugmills may also be produced locally. These pugmills have been used for centuries (33) and continue to prove successful. For example, they were in use in England a few decades ago(34), and more recently in Turkey(10). Indonesia may also start using these pugmills (27). Animal-powered pugmills (See figure IV.11) are made of a strong circular metal or wood tub, approximately 1 m high, with a vertical driven shaft in the centre, fitted with near-horizontal blades. Wet clay is driven downwards by the rotation of angled blades, gets cut and mixed by other blades and emerges from a small hole near the base of the tub. Animals are yoked to a beam which rotates the shaft (figure IV.12).

VI.2 Foot-treading

A simple and labour-intensive method of preparing the more plastic fine-grained clays is foot treading. Clay can be trodden in the quarry, but it is advisable to carry out this operation on a concrete or brickwork surface at the plant site (figure IV.13). Ideally, a circular or rectangular surface should be bounded by a 300 mm high upstand. The soil, straight from the quarry if suitable, or after some crushing, is spread 50 to 100 mm deep on the surface. If both fat and lean materials are needed, they should be layered. All the constituents must be thoroughly wetted, turned over with hoes or spades, and left covered up to temper for a few days. The workers then puddle the clay by treading in it, mixing and working the water in. The total volume of clay must be trodden systematically. It is advisable to turn it over with

Figure IV.10

Motor-driven pug mill (Madagascar)

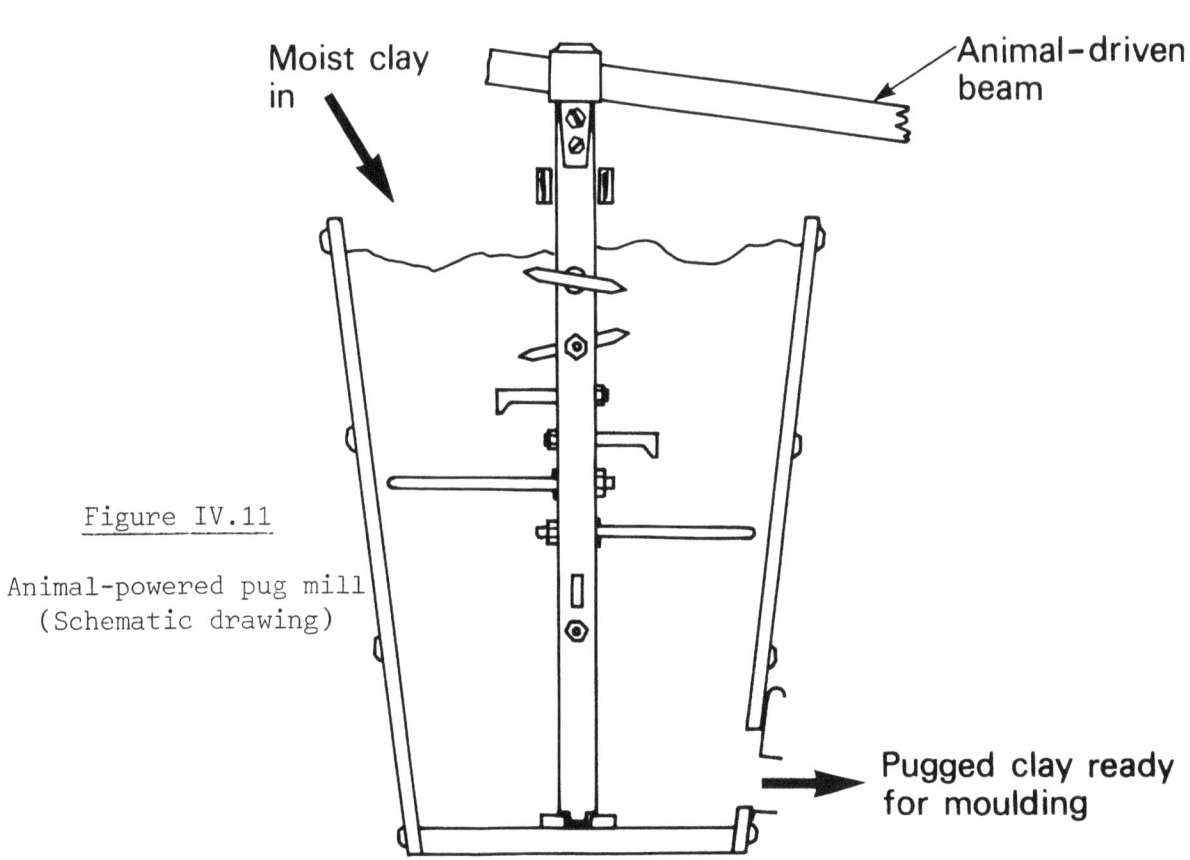

Figure IV.11

Animal-powered pug mill (Schematic drawing)

Figure IV.12

Animal-powered pug mill (Turkey)

Figure IV.13

Foot-treading clay on concrete floor (Ghana)

hoes and retread it. The mix should not be too dry as it will be difficult to move the feet up and down repeatedly. On the other hand,
if it is too wet, it will not be suitable for moulding. At best, foot treading is a very tiring work. If any treader feels a stone in the clay, he should pick it out, and discard it. Unfortunately, this is an operation which is easy to neglect and difficult to check. Fortunately, a few stones left in the clay will only damage a few bricks. Thus, for practical purposes, complete elimination of stones is not necessary for simple brick-moulding techniques. However, if bricks were to be wirecut (Chapter V), to include perforations, or to be made into extruded hollow blocks with thin walls the presence of stones would create problems.

In practice, it may be possible to get the workforce together just before the end of the working day in order to tread the wetted clay for the next or a later day's moulding. Leaving the clay for a while may further improve the moulding process.

VI.3 The washmill

The washmill is a labour-intensive method of cleaning clay in order to free it from stones, limestone nodules and other large particles. It requires fairly large quantities of water and produces a moulding material with a fairly high moisture content. The clay from the pit is preferably broken down prior to mixing with water in order to speed the process. A large brick-built or metal tank approximately 1 m deep and several metres across is filled to one-third of its height with the soil. Water is then added until it is two-thirds full. The mixture is stirred to disperse the soil up to the point where a clay slip or slurry is formed and the unwanted inclusions fall to the bottom of the tank. The clay slip, with sand suspended in it, is then run off into one of a series of lagoons or ponds where it will settle. After treatment of several loads of clay, the accumulated stones must be removed from the bottom of the washmill. Weeks may be required for the clay to settle in the lagoons, during which time the supernatent water can be run off a little at a time over a simple, variable-height sluice, and re-used in the washmill. Eventually, the last of the water will be drained off and the solid material will start to dry, though it must not dry completely. Once it can bear a person's weight, the material can be re-dug. Cuts should be taken from top to bottom to ensure an even mix of fine clay and sand at all times.

Washmills have been used with success in many places. One such washmill was operated at Bricket Wood, in the United Kingdom, prior to developments on the brickworks site by the Building Research Establishment. Recently, the washmill has solved a particular limestone problem in India(35), and is widely used at nearby Indore and other areas of Madhya Pradesh, where it is known as the 'ghol' method.

The shape of the washmill is not important if the slip is stirred manually. In this case it is easier to build a rectangular washmill. On the other hand, the use of other power sources requires a circular tank. Animal power could well be used for this purpose (figure IV.14). The size of mill will depend upon the nature and quantity of clay to be processed, the proportion of impurities and the planned brick production rate.

The washmill achieves most of the stages of clay preparation.

VII. TESTING

Testing methods described in Chapter II could be applied to check if the material is suitable or whether modifications need to be made. In practice, this information is difficult to apply for a given batch of clay since the latter must often be used before the test results are available. This emphasises the need for careful preparation to produce a material of constant properties, and the need to check the quality of the final product (Chapter VII) and to relate any problems to clay preparation.

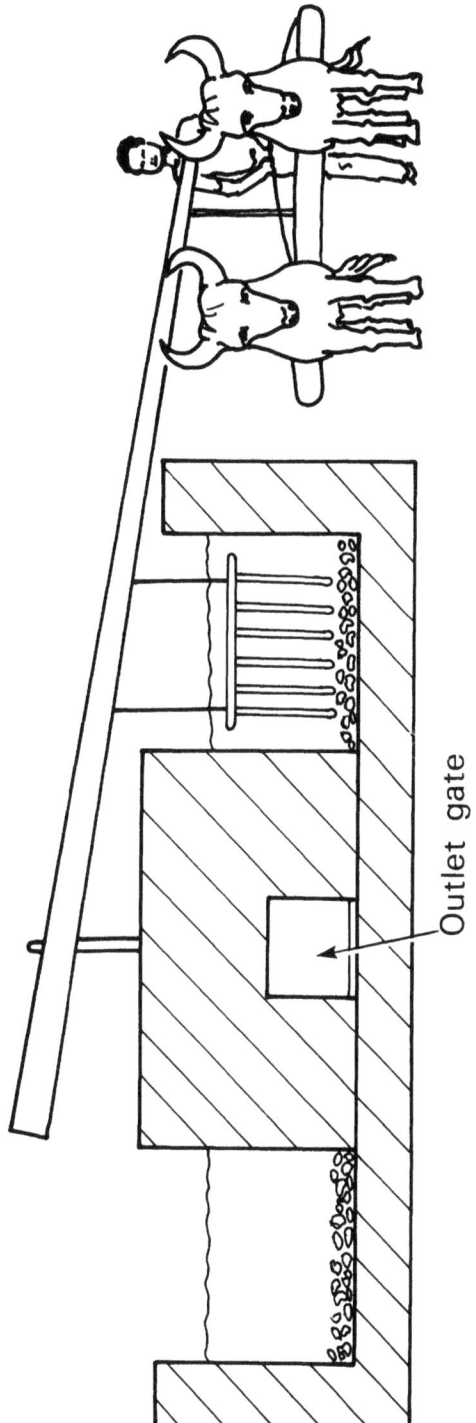

Figure IV.14

Animal-powered washmill

CHAPTER V

SHAPING

I. DESCRIPTION OF BRICKS TO BE PRODUCED

This section will provide the main characteristics of the bricks covered by this memorandum.

I.1 Size, shape and weight

In general, a brick should be of such a size and weight that it can be lifted in one hand. In almost all cases, a brick is twice as long as it is wide. Its height is usually one-third of its length (including the width of one mortar joint added to each end of the brick.) For example, the dimensions specified in the Indian Standard(36) are 190 x 90 x 90 mm and 190 x 90 x 40 mm. The 2:1 ratio is often adopted as it is highly suitable for bonding. The brick height is a matter of choice. In some countries, it is standardised so that it may fit to standardised dimensions of other components, such as window frames. The British Standard Specification for Bricks(37) requires that they be 215 x 102.5 x 65 mm or 225 x 112.5 x 75 mm if a 10 mm mortar joint is added. The 11-hole machine-extruded perforated brick (figure V.1) made in West Africa is too wide to be held in the hand. This type of brick may, however, be required for special purposes.

Bricks produced in small-scale plants do not generally have holes. In some cases, a frog is indented into one bed face. The single frog cannot be assumed to increase wall strength, but may assist in moulding in some circumstances. It may also shorten the drying time, slightly reduce the amount of fuel for firing, and reduce weight for transportation. However, it requires more mortar for wall building, especially if the brick is laid frog up. Frogs may be made on both bed faces in some shaping processes. Perforated bricks (figure V.1) are only produced by extrusion machines. They have advantages similar to those listed for frogged bricks.

Bricks which are accurate in shape and size are good to handle, transport, stack and build into a good wall with flat faces. If walls are to be rendered, less material is required than if they had an irregular surface. Furthermore, less mortar is required between accurately made bricks (see Chapter VIII).

Bricks of special shape may be produced, for example, for building wells and circular chimneys, or for joining walls of different thicknesses without sharp steps. Roofing tiles can also be made by similar methods. Big hollow blocks and decorative screen blocks are produced by machine extrusion. However, this memorandum concentrates on the production of ordinary-shaped bricks.

The size of the brick mould, die, etc. must be larger than the brick specification to allow for drying and firing shrinkages.

I.2 Faults in bricks

A number of faults in the finished product can be attributed to bad shaping. In extruded bricks, S-shaped cracks are caused by the clay-impeller's design or use. Saw tooth or dog-eared corners are caused by poor lubrication of the die near the corners. Internal cracks along the line of extrusion indicate an unequal extrusion rate in the centre as compared to that on the edge. Uneven heights of extruded bricks may be due to uneven spacing of cutting wires. Drag marks on cut surfaces are often due to dirty cutting wires in extruded and hand-made products. Irregular sizes and shapes of hand-moulded bricks may be due to inaccurate and bent moulds. Weaknesses may result from layers of sand being folded into the clay. Odd flashes may result from old clay stuck in narrow gaps of the mould or from overfilling of the mould. Missing corners, bent bricks, trapezoidal shapes and indentations may be due to incomplete filling of the mould, careless demoulding, setting down the green bricks sharply on the drying ground, squashing the demoulded bricks too tightly between pallets or marking them with the fingers.

Figure V.1

Machine-extruded and wire-cut bricks (Centre)
and hollow blocks (Ghana)

Figure V.2

Capital-intensive brickmaking factory (Ghana)

II. METHODS OF SHAPING

It has often been thought that machine-made bricks are better than hand-made bricks. This, however, should not be necessarily the case if hand-moulding is carried out with care. Furthermore, a comparative study of the economics of production in developing countries (see Chapter X) clearly indicates that hand-made production is still competitive in spite of technological developments.

Ready-made equipment may be imported, for both large-scale plants (e.g. equipment for a highly automated plant imported from Europe, by Ghana, shown in figure V.2) and small-scale units. In the latter case, equipment is used in a very few operations which complement the predominant use of labour.

Mechanised shaping methods will be considered briefly first, followed by the labour-intensive shaping methods. Finally, simple and cheap methods of assisting the hand-moulding process will be described.

The choice of shaping method should take into consideration the following: capital cost and expected life of equipment; maintenance and spares service; availability and cost of fuel (including reliability of electricity supply); scale of production in relation to raw materials supply; and market demand at time of installation and throughout the planned life of the installation. These points will be amplified in Chapters X and XI.

II.1 Mechanised shaping methods

II.1.1 Wire cut bricks

A method of producing machine-made bricks, which is commonly used in developing countries, is that of extrusion from an auger machine. This method is, for example, used in Madagascar (see figure V.3). In this machine, which is similar to a horizontal pugmill, the clay is impelled by an Archimedean screw. Taut wires cut brick sizes off the continuous column of clay (figure V.4 shows an automated cutter, manufactured in the Federal Republic of Germany and used in a brickmaking plant in Ghana).

This type of equipment is often imported by most developing countries. One exception is India where research was carried out into the construction of indigenous plants for making 20,000 bricks per day(38). Several such plants have since been built and operated(39). A supplier list of equipment is provided in Appendix IV. However, the wire cut-process is used at scales outside those considered in the memorandum.

Figure V.3

Extrusion machine (Madagascar)

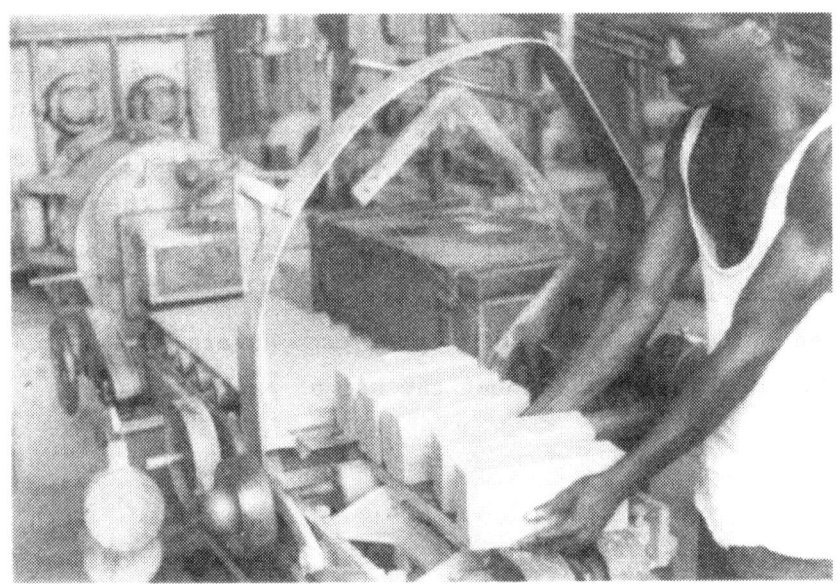

Figure V.4

Wire-cutter and wire-cut bricks (Ghana)

II.1.2 Soft mud process

Soft alluvial clays, such as those suitable for hand-moulding, may be processed by the soft-mud process. One of the smallest machines available produces approximately 14,000 bricks per day, a scale of production larger than those considered in this memorandum. This particular machine, originally made in the United Kingdom as the Berry Machine, was bought up by another company which is now producing it in the Netherlands (see Appendix IV). It has a horizontal pugmill followed by a set of cams which force moist clay through the side of the containing barrel into a quartet of iron-clad wooden moulds (see Figure V.5). To prevent clay from sticking, moulds are sanded. This can be done by hand. The whole process is fairly labour-intensive. The pugmill section alone could be used to prepare clays for hand-moulding.

II.1.3 Pressing

Bricks can be pressed, but commercially available machines are expensive and have high production rates. However, smaller hand-powered machines have been used in the past (33) and could still be employed, provided that the extra cost and time could be justified by an improvement in quality.

II.2 Hand-moulding

In the earliest techniques, soil was shaped by hand into lumps. The use of a mould to give shape to the soil resulted in more accurate and better bricks. Wood moulds should be soaked in oil for a few days before use. They are best made from a hard wood, shod on corners with iron or steel, and preferably lined with metal sheet. A handle bar is needed at each end. Metal moulds must be of a sufficient thickness to resist bending in use.

II.2.1 Slop-moulding

In slop-moulding, a very wet mix of clay is thrown into a wetted bottomless mould of wood or metal (figure V.6) as it rests upon a wooden pallet. Excess clay is scraped off either with the hand or with a striker (a straight wooden bar; see figure V.7). As the mix is very sloppy the mould cannot be removed until the brick has sufficiently dried or started to harden. Usually, the mould is removed immediately and returned to the moulding bench, re-wetted in a tub of water, and used again. In some instances, moulding is done on the ground without the use of pallet (40).

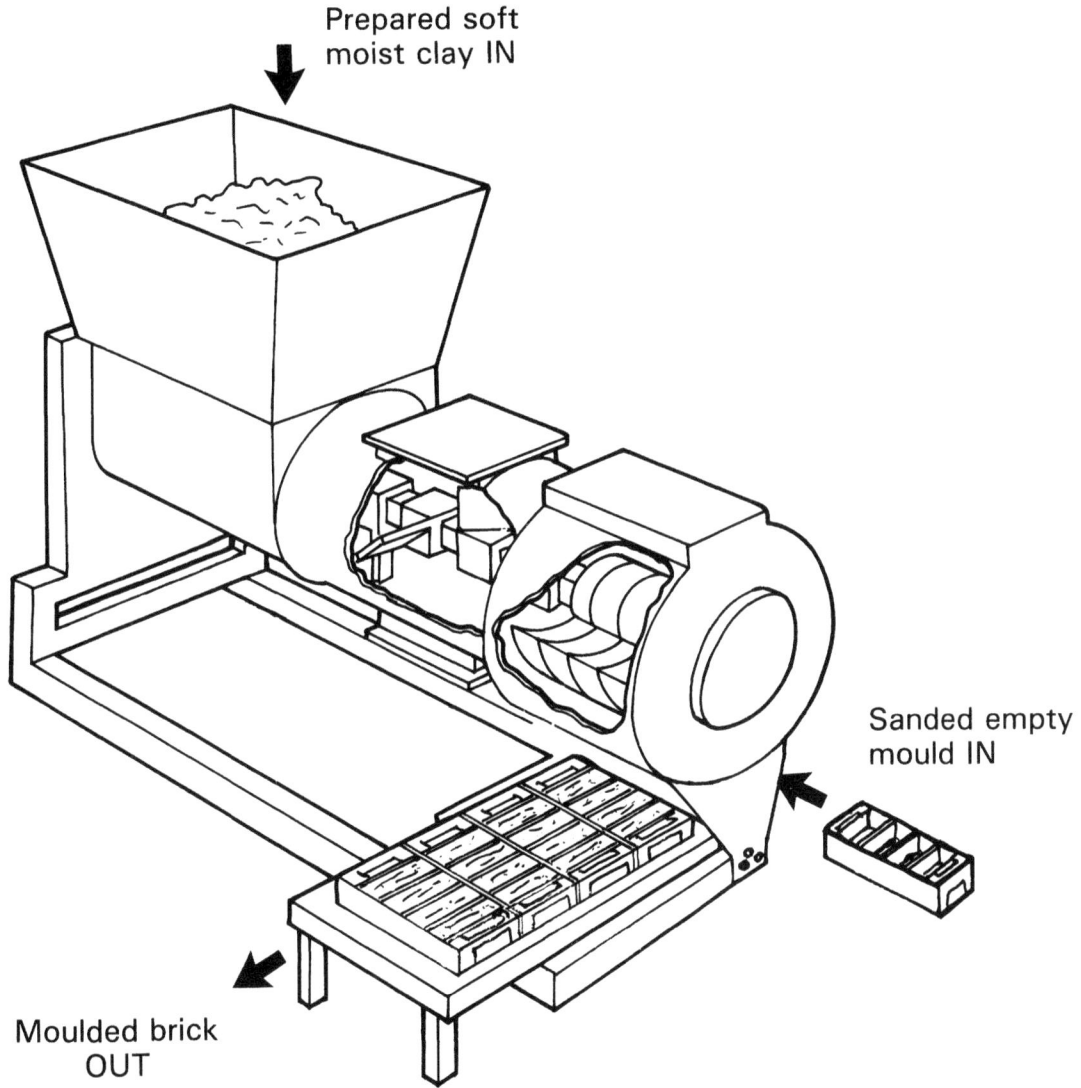

Figure V.5

Soft mud brickmoulding machine

Figure V.6

Slop-moulding in double cavity metal mould,
with moulder standing in deep hole (West Sudan)

Figure V.7

Striking off excess clay
in slop-moulding
(Ghana)

The slippery nature of the wet clay allows the demoulding of the brick. Two main disadvantages with this moulding method are distortion and high shrinkage.

The stickiness of the clay used in slop-moulding complicates the separation of the base of the mould from the brick. Some types of mould have bottoms fixed to the sides. Thus, a small gap between bottom and side is incorporated to allow air to pass in as the brick slides out.

The artisan brickmakers in Madagascar utilise the following method over the whole island. A basket of sand is kept alongside a tub of water (figure V.8). The sand is used to dust the base of the mould, thus preventing the clay from sticking to the base. These brickmakers carry their own bricks to the drying ground and invert the mould so that the sanded side is uppermost after demoulding. This practice is recommended as it minimises the difference between the evaporation rates of the top and bottom of the brick, thus avoiding distortions.

The brickmakers from Madagascar utilise another very simple moulding technique which is not normally used elsewhere, but would be worth adopting. A wooden plate, slightly smaller than the top of the brick, is laid on top of the latter. The thumbs are then placed on triangular upstands at the ends of the plate while the mould is being removed (see figure V.9). The plate is prevented from sticking to the brick by a layer of sand, and is left in position until required for the next demoulding operation. A maker's mark or trade mark can be moulded into the brick, using any design embossed on to the wooden plate.

II.2.2 Sand-moulding

Disadvantages of slop-moulding are partly overcome if a stiffer mix is used. One disadvantage of this is that the brick will not slip out of the mould easily. A layer of sand between clay and the mould surfaces will prevent sticking. The traditional moulder in India (figure V.10) sands the wet sides of the mould, then throws his clot of clay. The mould has a fixed bottom with a trade-mark on it. The bricks are finally demoulded at the drying ground (figure V.11). As can be seen from the figure, this method yields bricks with a regular shape and good finish, including the fine detail of the trade-mark in the frog of the bricks.

Figure V.8

Moulding with sand-covered mould base (Madagascar)

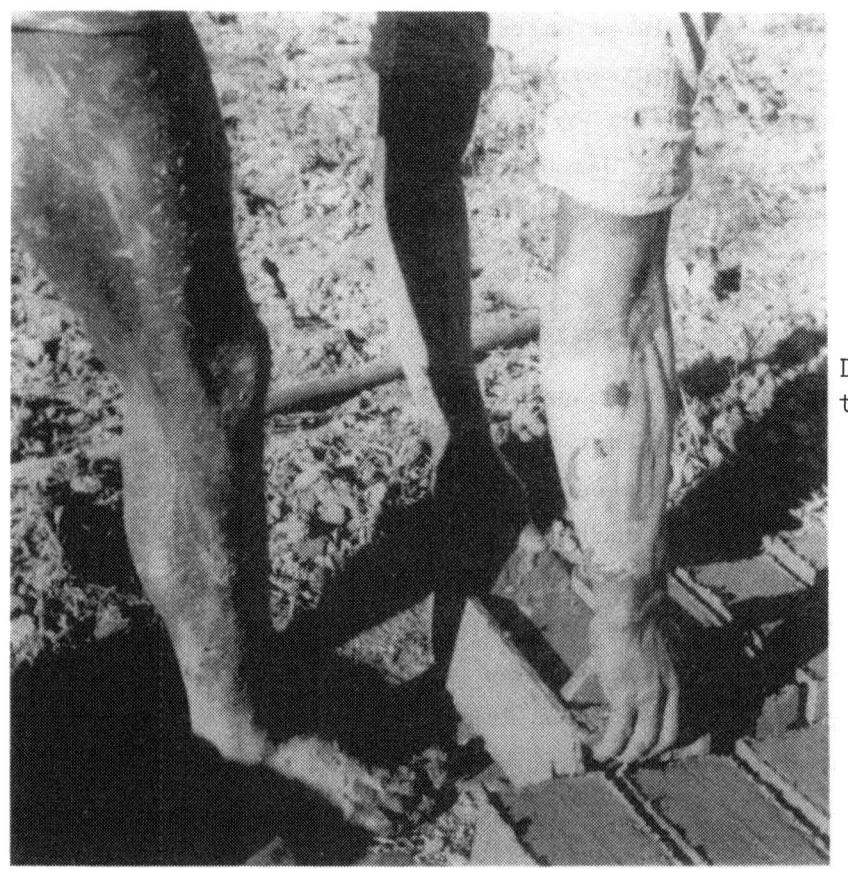

Figure V.9

Demoulding by pushing on triangular end pieces (Madagascar)

Figure V.10

Moulder using sand to prevent sticking (India)

Figure V.11

Two moulds demoulded at the drying ground. Good finish of produced bricks (India)

Attempts are often made to utilise the moulder's skills to the full, by having clot-makers prepare suitably sized pieces of clay ready for throwing into the mould. Elongated balls are prepared at a works in Ghana (figure V.12) and used in multi-cavity moulds in an attempt to increase productivity (figure V.13). In this works sawdust is used instead of sand to assist demoulding since it is readily available. The bricks are demoulded at the bench, on to a wooden pallet, then carried to the drying area five at a time.

Sand-moulding continues as a commercial method of production of hand-made bricks in the United Kingdom to satisfy a particular market requirement for bricks with variable texture and appearance. The moulders work at a bench with single cavity wood moulds. Instead of sanding the mould, the clot is covered in sand. Details of this method are illustrated in figure V.14. Given the stiffness of the mix, excess clay cannot be removed from the top of the mould by pushing with hands or a striker. Instead, a bow cutter, which consists of a taut wire on a bent stick or wooden frame, is used for this purpose.

Given the advantage of using stiffer clays, the Intermediate Technology Workshop (ITW) in the United Kingdom has developed a hinged bottom mould. Instead of having air inlet slits at the base, the air could be admitted by lifting the base from one end. This device has been successfuly used in commercial production in several countries in Africa (figure V.15), producing well-formed bricks.

Recently, the United Kingdom Building Research Establishment has commissioned the Intermediate Technology Workshop to develop hand-moulding processes to improve slop-moulding without the use of sand, since the latter is not available in all locations. One improvement proposed by ITW is the use of dry clay ground up finely to dust. The latter may then be used to prevent the clot from sticking in the mould, provided it remains in the mould for a few seconds. Quick demoulding is necessary as the dampness of the clot makes the dust sticky, and demoulding difficult.

ITW has developed two improved moulding devices which are described in a published document (41). A summary description of these two devices is provided below.

Figure V.12

Clot-makers rolling clay in sand (Ghana)

Figure V.13

Hinged-bottom mould and moulded brick (Southern Sudan)

Figure V.14

Moulder throws preformed clot. Use of five cavities mould (Ghana)

An essential part of good brick moulding is to dust the throwing clot on five of its six surfaces to prevent it sticking to the sides of the mould.

Experienced moulders develop their own individual techniques and the system illustrated is just one of many alternatives.

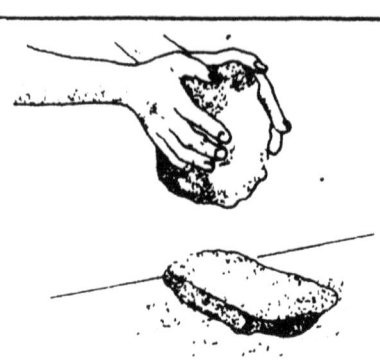

1. Starting with the cut off piece from the previous brick, put a new lump of clay on top.

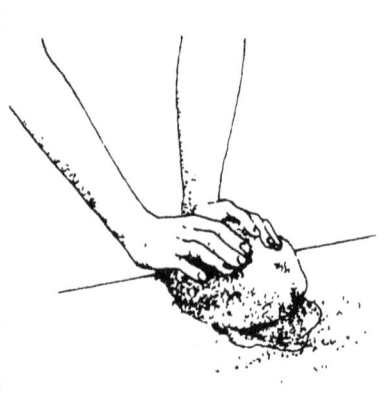

2. Using the 'heels' of the hand, roll the new ball of clay forward till the previous sanded cut-off comes over the top.

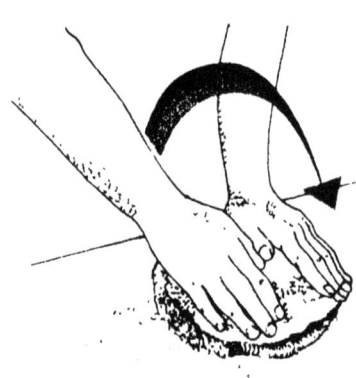

3. Press down making a wedge shape with sand now covering both large surfaces of the clot.

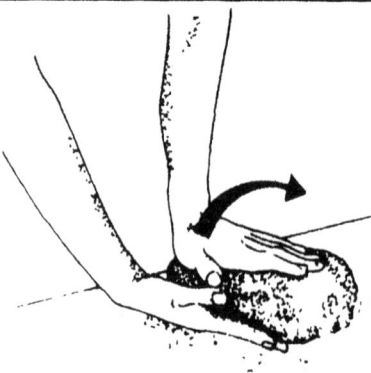

4. Lift the clot up on edge to sand the narrow edges.

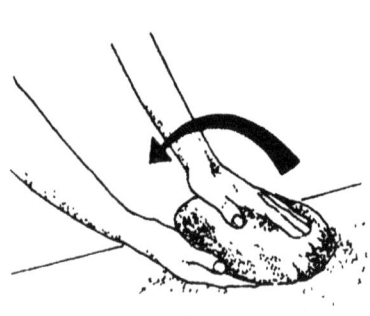

5. Apply sand to both edges so that now four sides are covered.

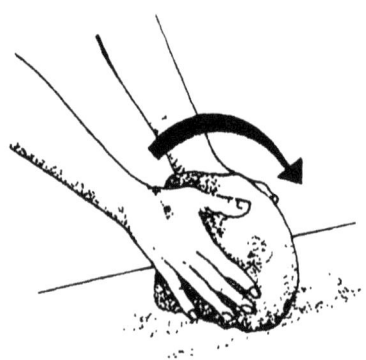

6. After sanding the second narrow edge, roll the clot forward to apply sand to the end.

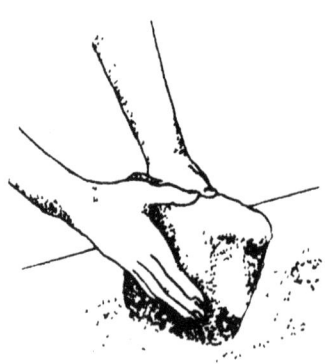

7. The clot should now be tapered with a wide top and narrow bottom end, ready to throw into the mould cavity.

Figure V.15

A technique for preparing a clot of clay

(i) *The turnover mould*

The turnover mould is another moulding device which may be easily produced locally. It is a four-sided wooden mould with a sheet steel lining turned in for 3 mm at the bottom to hold the clay in the mould. After the sanded clot is thrown and excess cut off, the four mould sides are lifted, freeing the moulded brick from the fixed base and its frog plate. The sides pivot on a simple bearing, and after a $180°$ rotation, a sharp knock on an end stop causes the moulded brick to drop out on to a pallet (which can be rotated with the mould). Figure V.16 illustrates the device and its operation.

(ii) *The table mould*

The local production of a second moulding device, the table mould, requires the skill of a carpenter or metalworker for the manufacture of some parts in order to ensure long reliable performance. Figure V.17 shows the mould recessed into the table. Once the sanded clot has been thrown and excess cut off with the built-in cutter, the brick may be ejected by raising the mould bottom by use of the foot-operated lever (figure V.18). Stiff bricks of excellent shape and clean arrises may be produced with this moulding device. It is advisable to transport them on light-weight hand-held wooden pallets in order to avoid finger-marking the smooth surface. The bricks are normally stiff enough to be stood on edge. Depending upon local skills and materials availability, the main structure of the table mould may be fabricated from box-section steel (figure V.18), steel angle and timber (figure V.17), or timber only (figure V.19). The mould should be lined with sheet steel and a frog plate may be fixed to the base. The cutter can either be built in or a separate bow cutter can be used instead. The sequence of moulding is illustrated in figure V.19.

Advice or supply of parts for this moulding device may be sought through the developers of the system, ITW. This moulding device has been used in several African countries, Sri Lanka and the Caribbean. A new feature which is being incorporated in the device consists of a lid which can be locked down on top of the mould, allowing the base to be raised slightly through a hand-operated lever and cam. This device allows some additional compaction of the clay, further improving the quality of the brick. However, it slows down production.

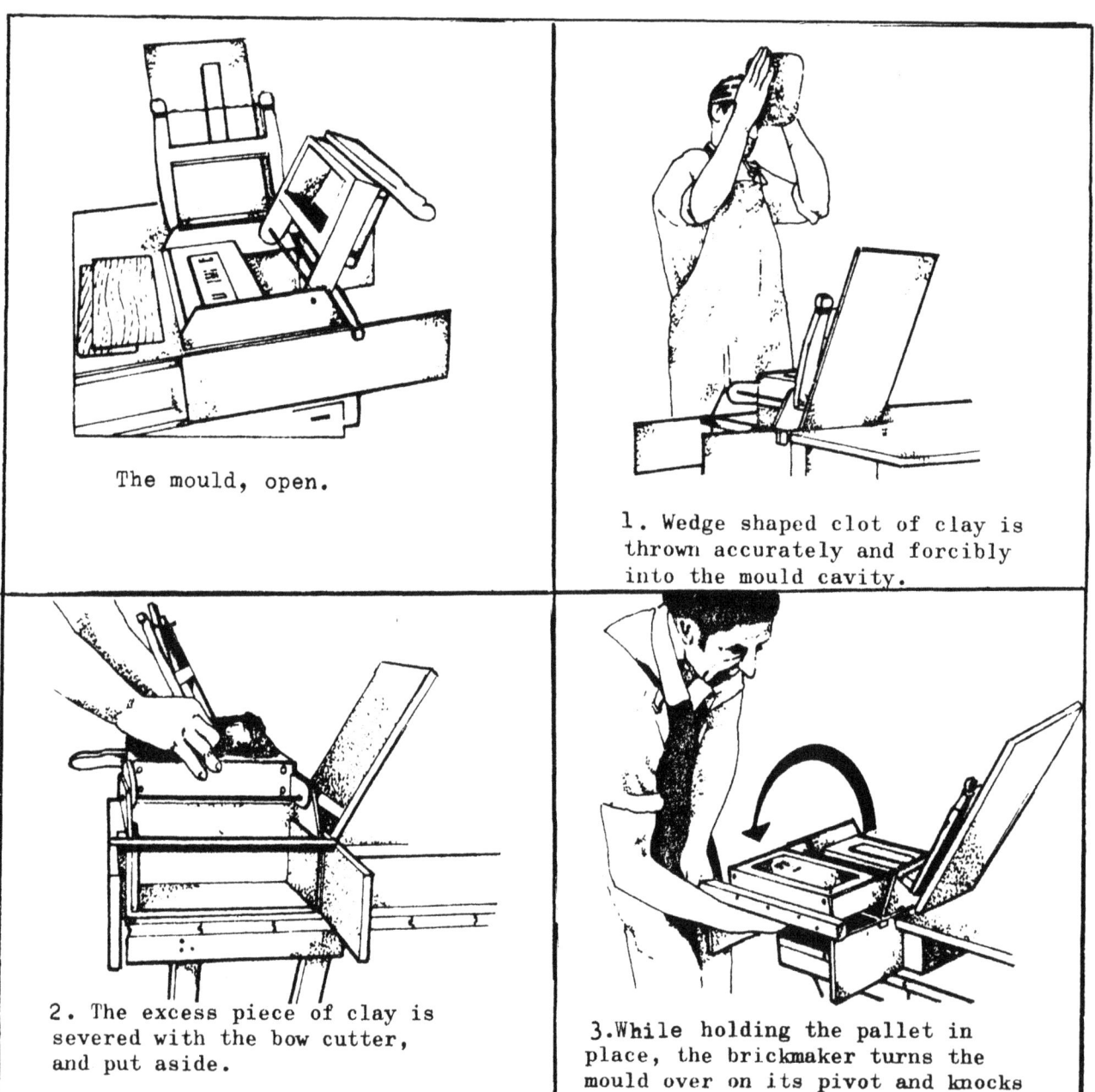

Figure V.16

The turnover mould and its operation

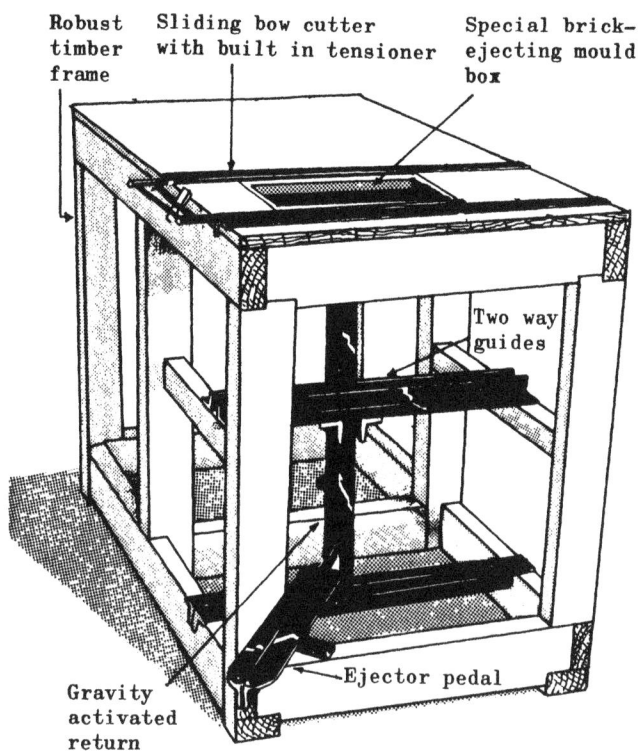

Figure V.17

The table mould

Figure V.18

Table mould; pedal being depressed to eject brick. Each brick is handled with pallets and stacked on racks (United Kingdom)

1. The wedge shaped clot of clay thrown forcibly into mould cavity.

2. At point of entry the bottom of the wedge of clay should come inside the rim of the mould on all sides to avoid 'cutting'.

3. After throwing the clot, a small mound of excess clay should be left on top.

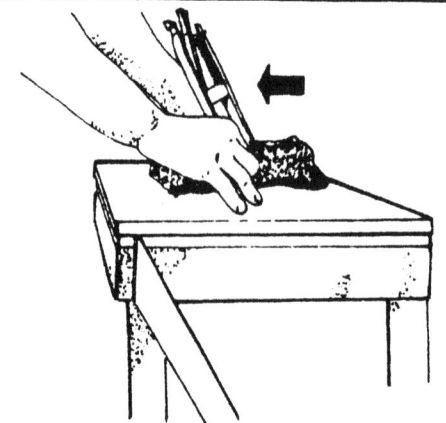

4. This excess clay is cut off with the bow cutter.

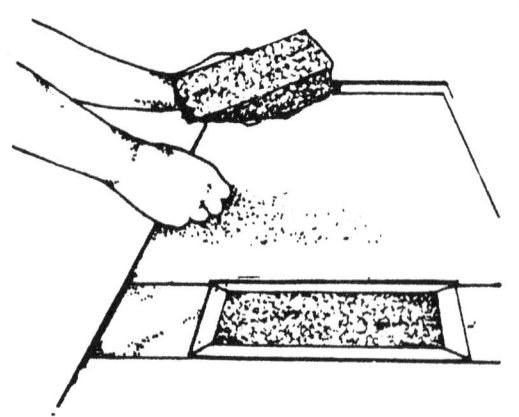

5. The off-cut is removed and set aside.

6. The pedal is pressed down to eject the brick.

Figure V.19

III. TRANSPORTATION OF BRICKS TO DRYING AREAS

Slop-moulding requires that bricks be carried in the mould to the drying area and set bed face down.

Where floors are fairly flat and hard, the labour needed to carry sand-moulded bricks, which are generally demoulded before transport, can be minimised by using barrows. ITW has developed racks for the transport of green bricks. Adapted hack barrows with forks are also available for the transport of 20 bricks at a time (figure V.18).

Sand-moulded bricks cannot be stacked on top of each other when first demoulded. Specially designed, large flat-top wheelbarrows may be used to carry 20 to 30 bricks at a time to the drying area. These bricks can be set on edge to dry rather than on the bed face. To minimise the load on the hands, the wheel of the barrow should be set well back under the load. Figure V.20 illustrates one hack barrow used through many years in the United Kingdom. Damage to bricks can be minimised if the wheel is sprung.

IV. SKILL REQUIREMENTS AND TRAINING

Mechanised brick-moulding requires a knowledge of clay characteristics as well as an in-depth understanding of mechanics and electricity. The mould-operators should also be able to undertake routine preventive maintenance and inspection for wear and tear of the equipment. They should also demonstrate an organisational ability in order to avoid loss of production through machine-servicing and repair. In practice, the machines will need constant adjustment during the working day.

Equipment used in hand-moulding should be strongly and accurately made, and kept clean. Hand-moulding skills require an appreciation of the nature of the clay and experience in throwing. In particular, the operator should be fully aware of the various remedial measures needed for the preparation of a mix with the right characteristics. He should also know how to throw with sufficient force and confidence so that the clot does not strike the side of the mould. Otherwise, a sanded clot would lose its sand and stick in the mould.

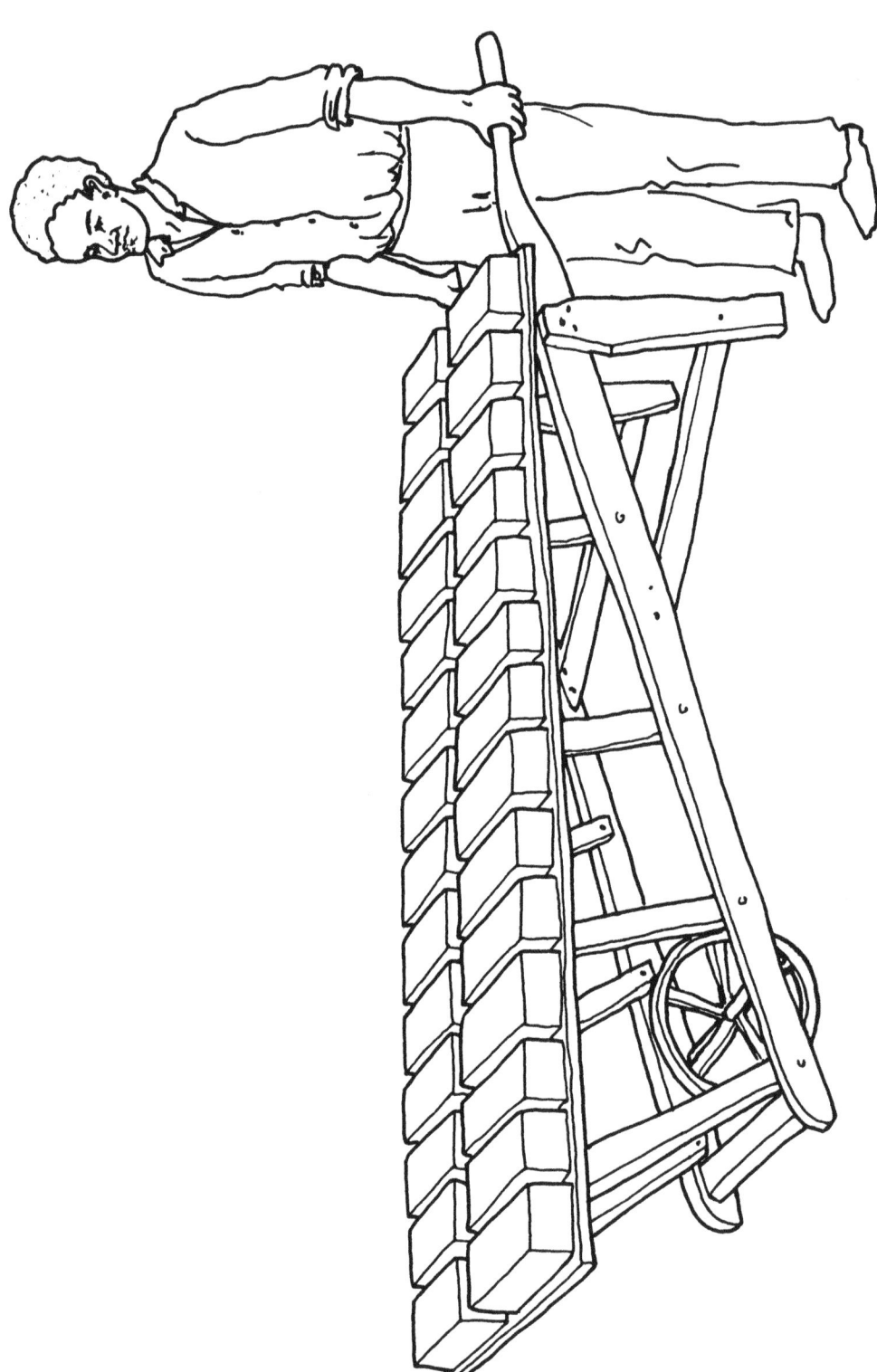

Figure V.20

Hack barrow

In general, skills are passed on within families. In addition, the techniques can be learnt elsewhere within a relatively short period of time. This is particularly the case for sand-moulding. There are a few 'tricks of the trade', which may help the beginner. Firstly, there is no substitute for a good hard accurate throw if the operator is to avoid the time-consuming pushing of clay into unfilled areas and the production of defective bricks. It is preferable to discard a poorly-moulded brick and send it back to the preparation stage than to go through subsequent stages and end up with bricks which cannot be marketed. Secondly, whenever sand-moulding is used with stiff clays, a hard bump of the mould on the bench helps to consolidate the clay and frees it from the mould. While the bumping and demoulding may be the most difficult skill to learn, such skill is not needed when the table mould is used. Thirdly, although the sand is the basic material preventing the clay from sticking, it is useful to wipe the surface of the mould occasionally with a rag soaked with oil in order to improve demoulding. The oil consumption is very small, and no special type of oil is needed. If none is available, a rag wetted with water may also be helpful. Little training is required for the sand-moulding process. For example, the well-formed and accurately-shaped brick in figure V.21 was moulded in a bottom-hinged mould by an inexperienced research scientist after only a few preliminary trials. The table mould seems eminently suitable when accurately-formed stiff bricks are required, where few skills exist and when little capital is available.

Figure V.21

Brick moulded in hinged-bottom mould
by non-craftsman

(United Kingdom)

V. PRODUCTIVITY OF LABOUR

It is not easy to measure or compare production rates as tasks are often shared. For example, clots may be made by the moulder himself or by a helper. Some moulders carry bricks to the drying area, while others have runners to perform the task. In mechanical production, it is difficult to assign a particular number of people to the shaping machine, since a man may be tending or servicing a variety of machines.

Table V.1 compares a number of mechanised and manual methods for the whole brickmaking process and for moulding only. Some of the estimates may include clay preparation, clot forming and carrying, while others may not.

Table V.1
Labour productivity in moulding
(man/hours per 1,000 bricks)

Area	Brick factory (complete process)		Area	Moulding only hand-made bricks
	Extruded wire cut bricks	Handmade bricks		
West Africa	37	97	East Africa	20
India	45	60	Madagascar	8
			United Kingdom (traditional)	21
			United Kingdom (table mould)	13

Source: 26, 33, 39, 41, 42.

Each country seems to have its own preferred method for hand-moulding: squatting and moulding on the ground in India, standing in Sudan, etc. In Madagascar, the moulder stands and works with his mould sloping away, fixed on top of a stout post. In the United Kingdom, a moulding bench is preferred. These methods are illustrated in some of the pictures within this chapter. These pictures also show that moulds may have one to five cavities. A simple cavity requires more running to the drying area for slop-moulding. On the other hand, five cavity moulds are very heavy when full. Changes to accepted methods of moulding would probably result in at least a temporary reduction in output. Where possible, however, it is recommended that the moulder stands rather than squats, uses the sand-moulding method and works on a table or bench.

A study of the brick industry in Colombia(43) concludes that direct evidence on the relative productivity of various technology levels is inadequate. However, indirect evidence indicates that it is worth while setting up labour-intensive units. The analysis carried out on the brick industry suggests that technologies imported from industrialised countries may not always be feasible.

Hand-moulding can produce bricks of technically good quality, at minimum capital cost. This labour-intensive method is highly suited to the varying market demand likely to be found in many developing countries. Some simple devices can lighten the burden and improve the working conditions of the brick moulders.

CHAPTER VI

DRYING

I. <u>OBJECTIVES OF DRYING</u>

There are several reasons for the drying of bricks before firing. These are briefly described below.

- In order to obtain the high strength and water-resistant properties of ceramic materials, the bricks must be burnt in a kiln to a high temperature. The bricks are piled up one on top of another, approximately 20 bricks high. Thus, the bricks at the bottom must be strong enough to carry the weight of those above. When first demoulded after shaping, the green brick may not be able to bear the weight of even one more brick without showing some distortion. When a certain amount of moisture has dried out, and the brick clay is approximately at the critical moisture content (see Chapter II), the bricks become "leather-hard". They are then sufficiently rigid and strong for handling and stacking.

- Once the "leather-hard" condition is reached, the bricks shrink. It is preferable that this shrinkage takes place before bricks are piled high for burning, lest the shrinkage causes the whole setting of bricks to become unstable, or to collapse within the kiln;

- Even after the "leather-hard" condition has been reached, there is much more water to be dried out of the bricks. If this is not done, the water in the bricks nearest to the heat source will evaporate and condense on cold bricks away from the heat source. These bricks will then absorb the water and get spoilt.

- Another risk is that water remaining in green bricks may turn to steam if the heat rises too quickly. This steam will build up pressure within the bricks, causing them to rupture. To minimise the risk, bricks should be as dry as possible before being put into the kiln;

- Within the kiln, any water remaining in the green bricks will only be driven out by burning expensive fuel. Fuel costs may thus be reduced if the maximum of water is removed through natural drying.

II. ARTIFICIAL DRYING

Drying should be completed with a minimum loss of bricks, and a minimum cracking and deformation of the latter (see Chapter II). The rate at which moisture evaporates from the surface should not be greater than the rate at which it can diffuse through the fine pores of the green brick. Thus, there is no purpose in creating more draught over the surface, or in heating the outside surface over a certain temperature. In fact, such action will cause a faster shrinkage of the surface than that of the interior of the brick, and thus cracking of the latter.

If drying is too slow, an opening material should be specified for the mix, or the brick should be reduced. A reduction in the volume of the brick may be achieved through the production of frogged or perforated bricks. The reduction of depth cannot, however, be carried too far if bricks were to be of a minimum strength. Once bricks are leather-hard, the drying rate can be increased.

If moisture diffuses to the surface, evaporates, but then remains just above the surface of the brick as a stagnant layer of moist air, further evaporation will be depressed. This will happen even when bricks are heated in a dryer. In this case, the drying process should be modified in order to avoid the above problem.

The rate of drying depends on diffusion rate of moisture in brick, temperature, humidity and air speed. In temperate climates, where drying is not possible during the cold damp winter months, medium brick plants use floors heated by fires. This raises the air temperature and reduces the relative humidity, thus enabling drying to take place. This drying is costly as it requires large amounts of fuel. Other systems include chamber dryers for batches of bricks and tunnel dryers which operate continuously.

In the operation of artificial dryers, energy is required to supply the latent heat of evaporation of the water as well as to heat the bricks and the air passing through the dryer. It is also required for other heat losses into the surroundings.

Estimates of total energy requirements for evaporating one kilogram of water are provided in Table VI.1 below:

Table VI.1

Energy requirements for drying bricks

Efficiency of process	Energy (kJ per kg water evaporated)		
	Hot floor dryer	Chamber dryer	Tunnel dryer
High	7 100	3 300	3 300
Low	12 400	8 900	7 100

Source: 45

Calculated energy requirements for the drying of bricks range from 3,000 to 8,000 MJ per 1,000 bricks. These energy requirements are relatively important when compared to those needed for the firing of bricks (i.e. 5,000 to 16,000 MJ per 1,000 bricks(8, 10)). Hence, it should be realised that energy for artificial drying is of similar magnitude to that required for firing.

Artificial drying is mostly needed in the automated large-scale works, where automatic handling of bricks might be justified economically, and where kiln designs may be such that waste heat can be conducted from kilns to dryers. The tunnel kiln lends itself to recovery of heat from the cooling zone, but in less sophisticated and smaller kilns it is difficult and relatively expensive to collect the heat for drying. One cheap alternative could be to dry bricks by placing them around the kiln. This alternative may not be feasible if space is not sufficient for drying all the bricks, and if it were to restrict access to the kiln.

A recent drying method involves the use of rotating air dryers. In these dryers, hot air is applied as occasional blasts along strategically positioned racks full of bricks, the draught removing the accumulated layer of moist air which has collected close to the brick surface.

In some tropical countries, drying is difficult during the wet season. Consequently, large plants are often equipped with air dryers. Figure VI.1 shows rotary dryers awaiting installation in a Ghanaian brickwork.

Bricks can be weighed from time to time to estimate the amount of water which has dried out. These estimates may then be used to determine the time needed for bricks to reach the "leather-hard" stage.

In view of the high cost and/or the scarcity of fuel, the hot weather conditions in a large number of countries, overall production considerations and market requirements, the use of artificial drying does not seem appropriate for small-scale brickmaking in most developing countries.

III. NATURAL DRYING

The main advantage of natural drying over artificial drying is the saving of fuel. The principal disadvantage is the amount of extra handling which may be necessary, resulting in higher labour costs and a greater risk of damage to bricks.

The two main methods of natural drying are (i) fully exposed or temporarily covered bricks and (ii) bricks drying under a shed with a permanent roof.

III.1 Drying on the ground and in hacks

Slop-moulded bricks must be dried in single layers as they cannot bear the weight of other bricks without distorting. They should be laid as close as possible to economise on the use of drying area (figure VI.2). Approximately one-tenth of a hectare should be allowed for every thousand bricks produced per day. The ground should be swept free of loose debris as the latter will be picked up by the underneath surface of the soft brick (figure VI.3). It should also be as flat as possible since any bumps or dents, even the small ones shown in figure VI.2, will not support the soft brick properly and will tend to distort it (figure VI.3). A thin layer of sand may be spread over a bumpy ground in order to improve drying conditions (figure VI.4).

Figure VI.1

Rotary dryers (Ghana)

Figure VI.2

Bumpy drying ground covered with debris (Madagascar)

Figure VI.3

Partly dried brick distorted and
covered with debris (West Sudan)

Figure VI.4

Good bricks dried on clean, level ground
(West Sudan)

Sand-moulded bricks, being more firm than slop-moulded bricks, are not likely to pick up debris or to distort easily. They may also be set down on edge, which further reduces the chance of distortion due to uneven support.

As soon as bricks are dry enough to handle (i.e. once they have reached the leather-hard stage), they should be turned right over to allow the face which was in contact with the ground to dry. The required drying time will vary with weather conditions, but should not generally exceed one or two days.

After approximately three more days, depending upon weather conditions, the bricks can be removed from the drying ground and built into long open work walls eight or ten bricks high. These walls are known as hacks (figure VI.5). The drying of bricks should continue for a few weeks more. Bricks become lighter in colour as they dry. Sample bricks should be broken in order to check whether the inside is of light colour and therefore dry.

The bottom bricks in the hack may remain wet if laid directly on the ground. It is therefore advisable to build the hacks on already burnt bricks or wooden planks, which may be left in position permanently.

Although light rain may not harm exposed green bricks, heavy rain can have serious consequences, and days of hard work may be wasted in a few minutes. Figure VI.6 shows bricks damaged by rain and bricks which are laid out for drying after the rain storms. Where the risk of heavy rains exists it is almost impossible to carry all the bricks under cover, especially newly-made slop-moulded ones. Instead, one may need to use plastic sheeting for protection against the rain. Plastic sheeting may not, however, give the best protection as it is difficult to lay the plastic over a large amount of bricks unless the stacks of bricks are separated by walkways. Furthermore, the plastic sheets must be weighted down to prevent the wind from blowing them away. Since bricks close to the edges of the plastic sheets may not be protected effectively, it is advisable to use special cover or permanent sheds whenever drying takes place during uncertain weather conditions.

One of the chief reasons for the cracking and distortion of bricks during the drying stage is the high rate of drying. This rate is difficult to control in the open air, although the use of leaves or grass on top of the bricks may help control drying to some extent. A high rate of drying may result from the action of the sun as well as from low humidity levels and winds. Thus, it may take place even when the bricks are shaded.

Figure VI.5

Drying being completed in hacks
(Madagascar)

Figure VI.6

Rain damage to green bricks - No damage on
protected bricks

Hack drying is an effective drying method used in the temperate climate of the United Kingdom over many years. In this method, bricks are transported bed face down on the hack barrow and set on edge on wooden planks in two long parallel rows. Hack covers, made from pairs of thin sheets of wood tied with a cross bar long enough to span both rows of bricks, are laid carefully on the green bricks for protection against rain. By the time the end of the row is reached, a second course of green bricks may be laid without distorting those in the first row. A maximum of six courses of bricks may be built in this manner. Such a hack ground is illustrated in figure VI.7. The boards are kept at the correct separation by an occasional batten fastened between them. Only the tie-bars of the covers rest on the bricks. Thus, the sloping covers do not damage the upper corners of the bricks. These covers are sufficient to protect bricks from the heavy vertically falling rain. In rainy and windy weather an extra board or sacking may be placed on the side facing the wind. Rainwater from the edge of the roof falls into a gulley on either side of the hack. The earth dug from the gulley is spread between gullies in order to raise the plank slightly. The preferred direction of the hacks should be such that the sun shines on either side of the wall for equal periods of time. Figure VI.8 shows bricks close-set in a covered hack. When they are dry enough to handle, they can be reset with wider spacings, the hacks then becoming slightly higher. Wider spacing increases the drying rate, which may be necessary after the leather-hard stage has been reached. Hacks must be sufficiently apart to allow access and operation. The drying of bricks with this method takes several weeks. An area for the hack ground of approximately one-tenth of a hectare should be allocated for an output of 1,000 bricks per day(8). This drying method may be adopted in showery weather or during short periods of rain in many countries.

III.2 **Drying in a shed**

Provision of a shed, sufficiently large to cover both the moulding and drying areas and, possibly, the kiln, may be considered in some cases. Such a shed may, however, constitute the largest item of capital expenditure in setting up a small brickworks. A shed covering only the drying grounds will be cheaper and will improve the chances of successful drying. Figure VI.9 shows a simple structure used in Madagascar. This shed may be used efficiently if bricks can be built up in hacks when firm enough.

Figure VI.7

Transferring freshly moulded bricks from barrow to dry in hacks with moveable covers

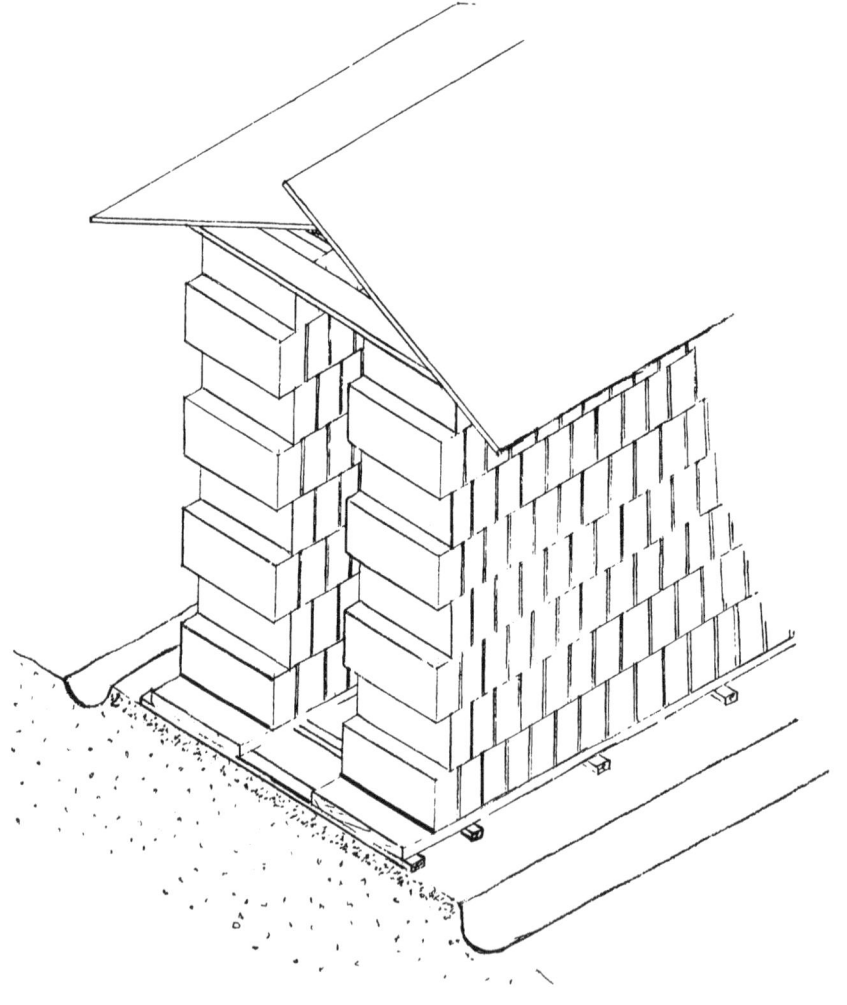

Figure VI.8

Covered drying hack

The use of racks reduces the volume needed for the drying of a given number of bricks (see figure VI.10). Such racks enable soft, freshly moulded bricks to dry one brick high on each shelf. Thus, they will not be distorted by other bricks above them. Some cracking may take place in bricks nearest the outside of the shed, where direct sun, low humidity and great air movement cause faster drying. Side screens are necessary under such circumstances. Bricks should be turned on to their sides once they have reached the leather-hard stage after a few days of drying (28).

Side screens allow a better control of the drying procedure. They may be made out of plastic, cloth, wood, metal, woven grass or bamboo. The most convenient side screens should be of a type which can be rolled up easily when not required. When down, they keep off driving rain and direct sunshine, increase and equalise humidity within the shed, and prevent excessive air circulation through the stacks of bricks. Access to the shed should be from the side sheltered from the wind. When drying is too slow, screens can be raised.

A cheaply constructed but effective drying shed is shown in figure VI.11. Pole timbers support a roof, high enough to allow workers to operate within the shed. Rudimentary but effective side screens are available for rapid use. Racks for holding the green blocks consist of planks, supported at the ends on previously-fired bricks, thus avoiding the necessity for substantial wooden posts to bear the weight of the loaded shelves. For safety reasons, these racks must not wobble or tilt.

IV. SHRINKAGE

Shrinkage is inevitable on drying clayware such as bricks. The most important rule is to dry bricks as slowly as possible in order to minimise stresses and the incidence of cracking and distortion. A 7 per cent linear shrinkage should not cause difficulties in subsequent processing. A 10 per cent linear shrinkage may also be acceptable with some clays if drying is carried out carefully.

Problems may be lessened if the clay proportion in the mix is reduced by the addition of sand or grog.

Fully shrunk bricks are not completely dry. Further drying is needed before they are ready for firing in the kiln.

Figure VI.9

Open-sided drying shed (Madagascar)

Figure VI.10

Rack-drying under cover (Madagascar)

Figure VI.11

Rack-drying under cover, with removeable side screens (Madagascar)

CHAPTER VII

FIRING

I. <u>OBJECTIVES OF FIRING</u>

The firing of green bricks changes their physical structure and gives them good mechanical properties and resistance to slaking by water. If carried out properly, the firing process should minimise the occurrence of the following problems:

- the splitting of bricks due to the incomplete removal of moisture before firing;

- Low strength bricks due to insufficiently hard firing;

- Slaking by water, due to inadequate control of the firing temperature;

- Bricks fused together, melted on one face, or distorted by the load imposed by other bricks on top; these problems are caused by too high a temperature;

- Variety of sizes of fired bricks although the green bricks were of the same size; this is caused by temperature variations between different parts of the kiln;

- Fine cracking over brick surfaces resulting from a too rapid temperature change, either during heating or cooling, or from condensation of water vapour from heated bricks on to cooler bricks(47);

- Local cracking over hard lumps or stones mixed in the clay. These inclusions should have been removed during clay preparation, although rapid changes of temperature may have aggravated the problem;

- Black cores in bricks: these are not necessarily detrimental if they are due to the presence of carbon although the latter ought to have been burnt up as a contributory fuel in the interest of cost reduction (48). Black cores may also be due to iron in the reduced, low valency ferrous form. Both causes may be remedied by providing sufficient oxygen (e.g. by leaving adequate paths for the flow of air between the bricks and through the kiln(47)). An opening material may also need to be mixed in the body;

- Bloating: cracked blisters appear on the surface of the bricks as a result of the pressure of gases produced after vitrification has commenced; holding the temperature steady at an earlier stage could allow the gases to diffuse out while the body is still permeable. The problem may also be alleviated by incorporating an opening material in the clay;

- Limebursting: this problem may be solved by removal or grinding of pieces of limestone, or in some cases, by adding salt(49). Alternatively, the quicklime formed within the brick may be dead burnt by heating to approximately $1,100^0 C$ (50);

- Efflorescence and sulphate attack of cement-based mortars and renderings: this problem may be reduced by harder firing, as an alternative to other methods and precautions (see Chapter IV);

- Scum on brick surfaces may be minimised by preventing condensation of products of combustion on cold, green bricks.

II. TECHNIQUES OF FIRING

Whatever system of firing is used, it is recommended that a low heat be first applied to the green bricks in order to drive off any residual moisture. This should continue until no more steam is evolved. This part of the heating is known as water-smoking. The completion of this firing stage may be simply tested by inserting a cold iron bar into a space purposely left between the bricks in the kiln, and by withdrawing it after only a few seconds. Condensation on the bar indicates that steam is still being evolved, and low heat should be continued until no condensation is found on the cold bar after re-insertion. This could take a whole week in some cases and needs plenty of air through the kiln.

Once water-smoking is complete, a rate of rise of temperature of 50^0C per hour may be safe in fully-controlled kilns, which are outside the scope of this memorandum. At the critical temperatures (e.g. quartz inversion) the rate can be temporarily reduced to avoid problems. In more simple kilns, the heating rate should be slower partly because of the lack of precise temperature control and partly because of the impossibility of getting enough fuel burning in some kiln designs. Although a slow rate of heating is safer, faster rates involve less heating time, lower heat losses and, therefore, lower costs. The optimal heating rate is that which requires the shortest heating schedule while yielding a product of satisfactory quality. A maximum of two weeks may be needed for the whole firing process.

Maximum temperatures with little air should be held for at least several hours. A whole day may be needed with some kilns with poor heat distribution in order to ensure a maximum yield of good quality bricks. During this firing stage, known as the "soaking" stage, the heat diffuses through the kiln, various chemical reactions take place and a glassy material is formed. Once the soaking is complete, the heat source may be removed.

The cooling rate should not be too rapid. In practice, natural cooling within the large mass of thousands of bricks, with limited air flow, is satisfactory. Cooling may take a whole week. More air may be allowed in once lower temperatures are reached in order to speed up cooling.

III. KILN DESIGNS

In general, large kilns are more economical on the use of fuel than small kilns as less heat is lost through the proportionally smaller outside area of the kiln. Thus, separate teams of brickmakers may use the same large kiln on a co-operative basis, and thus benefit from lower firing costs.

There is a wide variety of kiln types and sizes. These may be split into two major groups: the intermittent kilns and the continuous kilns.

Intermittent kilns are filled with green bricks which are first heated up to the maximum temperature and then cooled before they are drawn out from the kiln. Thus, the kiln structure is also heated during the process. Consequently, all the heat within the bricks and kiln is lost into the atmosphere during cooling. Intermittent kilns are very adaptable to changing market demands, but are not the most fuel efficient. They include the clamp, scove, scotch and downdraught kilns.

The continuous kilns have fires alight in some part of them all the time. Fired bricks are continuously removed and replaced by green bricks in another part of the kiln which is then heated. Consequently, the rate of output is approximately constant. The continuous group of kilns includes various versions of the Hoffmann kiln, including the Bull's trench, zig-zag and high draft kilns, and the tunnel kiln. The latter is a capital-intensive, large-scale continuous kiln, which is outside the scope of this memorandum.[1] Continuous kilns utilise heat from the cooling bricks to pre-heat green bricks and combustion air, or to dry bricks before they are put into the kiln. Consequently, continuous kilns are economical in the use of fuel.

III.1 The clamp

The clamp is the most basic type of kiln since no permanent kiln structure is built. It consists essentially of a pile of green bricks interspersed with combustible material. Normally, the clay from which the bricks are moulded also includes fuel material. The clamp kilns were commonly used in the United Kingdom. Some of these, containing one and a quarter million bricks, are still used for the production of bricks of various colours and textures.

It is possible to use a variety of burnable waste materials in brick clays (e.g. sifted rubbish, small particles of coke, coal dust with ashes, breeze). In countries where timber is produced, large quantities of sawdust may be mixed with clay before firing. This will reduce the expenditure on the main fuel for burning the bricks. Waste materials should be of relatively small size and should not exceed in weight 5 to 10 per cent of the total mixture. Otherwise, the clay will become difficult to mould or the finished product may become too weak or too porous. Furthermore, the added fuel material should be thoroughly mixed with the clay.

A flat, dry area of land is first chosen, and a checkerwork pattern of spaced out, already burnt bricks laid down over an area of approximately 15 m by 12 m. Fuel in the form of coke, breeze or small coal[2] is then spread between the checkerwork bricks, covering the latter with a layer at least 20 cm thick. Dry, green bricks are next closed-laid on edge upon this fuel bed.

---

[1] In the tunnel kiln, bricks stacked on heat-resistant trolleys or cars are subjected to increasingly hotter temperatures, then cool off before leaving the tunnel.

[2] Coke or breeze is generally used in clamp kilns in the United Kingdom. Small coal is used in a number of developing countries, such as Zambia (51).

A clamp is generally made up of approximately 28 layers of bricks. Its sides are sloped for stability (see figures VII.1 and VII.2). Three or four holes[1] at the base of one of the clamp walls are formed in order to allow the initial ignition of the fuel bed. Two courses of already fired bricks are next laid on top of the green bricks for insulation purposes. Fired bricks are also laid against the sloping sides of the clamp as it progresses. Sometimes, a second thin bed of fuel is laid at a higher level in the clamp(51).

Once several metres of the length of the clamp have been built up, the fuel bed may be ignited with wood stuffed into the eyes of the clamp. The latter are bricked up with loosely placed burnt bricks once the fuel bed is alight. As the fire advances, more green bricks are built into the clamp.

During burning, the heat rises through the bricks above, fumes and sometimes smoke leaving the top of the clamp. The rate of burning is not easily controlled and depends upon several factors, including wind strength and direction. Some wind protection with screens can help control the temperature. Ventilation, and hence burning rate, can also be controlled, to some extent by an adjustment of the burnt brick covering the top of the clamp. For example, these bricks may be spaced out or removed in order to speed up the firing of the bricks in a given area. Conversely, they may be tightened up or covered with ashes in order to slow down the burning rate. It is desirable to have the fire advancing with a straight front at a steady rate. Bricks close to the edges of the clamp will tend to be underfired as a result of higher heat losses. This may be partially rectified by placing a little more fuel near the edges of the clamp. Extra fuel may also be spread between the top bricks during firing.

The firing process is indicated by the sinking of the top of the clamp. Under the right circumstances, the latter will settle evenly. Once the fire has passed through a particular point, the bricks start to cool. They may then be withdrawn, sorted into various grades, and sold. Thus, bricks within one clamp are set and drawn simultaneously. After a number of weeks, the fire reaches the end of the clamp. Before then, construction and lighting of a new clamp may be started as previously, if market demand requires it.

---

[1] These holes are known as the "eyes" of the clamp.

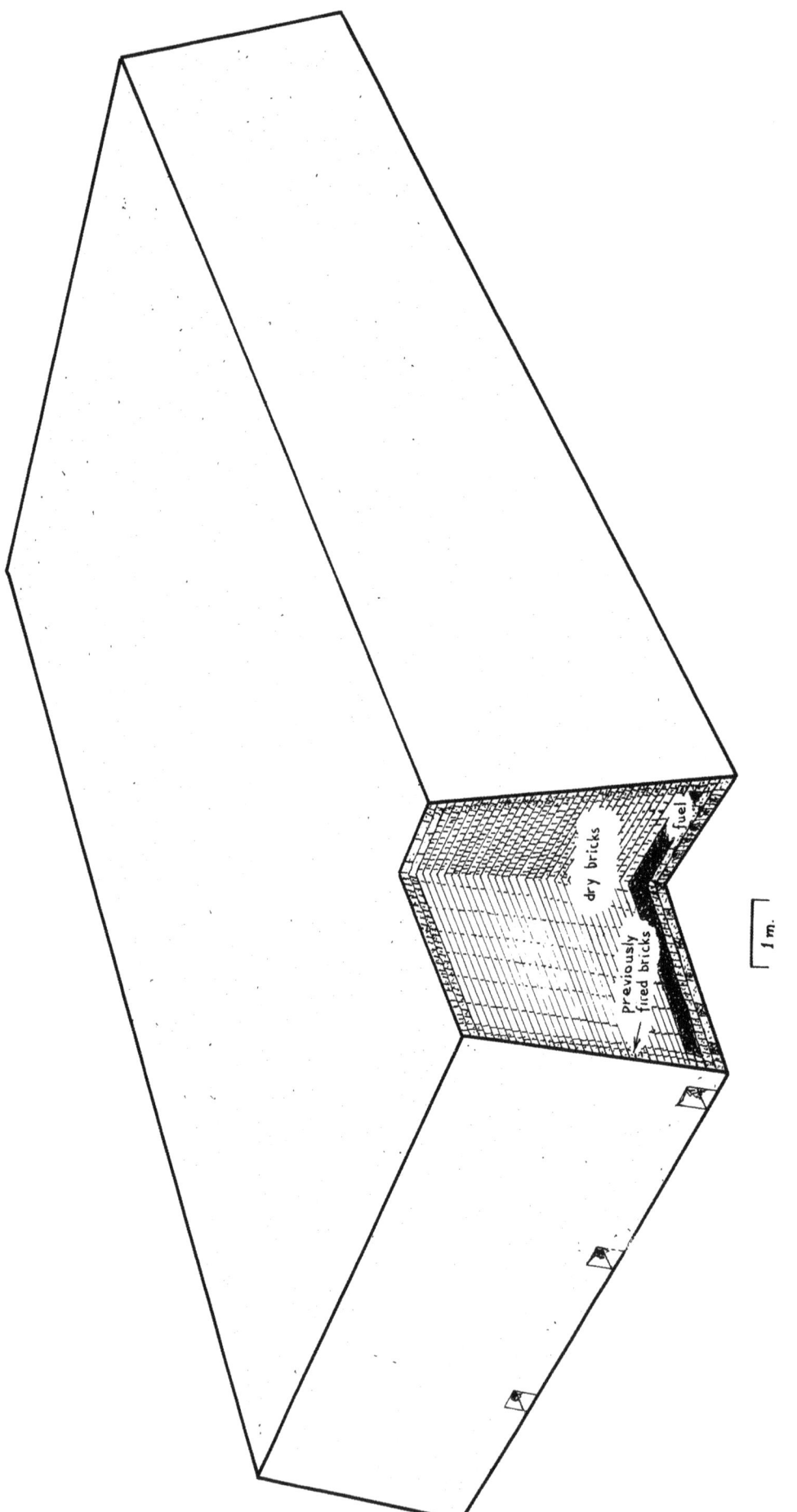

Figure VII.1

Clamp kiln - schematic drawing

Figure VII.2

Clamp kiln (United Kingdom)

If enough air flows through the bricks during firing, the oxidising process will give them a red colour. Where air is scarce, reducing conditions due to the gases from the burnt fuel will yield orange or yellow bricks, especially if a limy clay is used for moulding. Variations in colours will be normal even on a single brick face.

As the fuel is in close contact with the green bricks, the fuel efficiency of a large clamp of 100,000 to 1 million bricks can be fairly high (e.g. about 7,000 MJ per 1,000 bricks). Smaller clamps will be less efficient, as a result of the greater proportion of outer cooling area for a given volume. However, they may be operated successfully with only 10,000 bricks. For low production rates, it is only necessary to fire a clamp occasionally and have the bricks in place until they are sold.

The bricks near the centre of the clamp will be the hardest. Others should be sufficiently good for many uses. They should be sorted for sale as best, "seconds" and soft-burnt bricks. However, 20 per cent of the bricks may still not be saleable. Fortunately, many of these rejects can be put into the next clamp for refiring, or used in the clamp base, sides or top.

III.2 The scove kiln

A widely used adaptation of the clamp is the scove kiln, also mistakenly called a clamp. If the fuel available is of a type which cannot be spread as a thin bed at the base of the kiln and/or is not in sufficient quantity to burn all the bricks without the need for replenishment, tunnels can be built through the base of the pile in order to feed additional fuel (figure VII.3). This is a suitable method of burning wood, the latter being one of the most frequently used fuels for small-scale brickmaking in developing countries. Usually, the outer surface of the piled-up bricks is scoved, that is to say plastered all over its sides, with mud (figure VII.4). Thus, the name of the scove kiln.

The construction of a scove requires a level, dry area of land. Previously fired bricks, if available, lay bed face down to form a good, flat surface. Three or four layers of bricks are used to form the bottom of the tunnels. The width of each tunnel is approximately equal to that of two brick lengths. Three lengths of bricks separate the tunnels. Alternate courses are laid at right angles to each other (i.e. a course of headers, followed by a course of stretchers). Two short tunnels (e.g. approximately 2 m long) may be

Figure VII.3

Scove (Madagascar)

Figure VII.4

Scoving face of kiln (Madagascar)

sufficient for a small number of bricks. For large numbers of bricks, tunnels cannot be longer than approximately 6 m Otherwise, fuel inserted from both ends will not reach the centre of the tunnel. Large numbers of bricks are dealt with by extending the number of tunnels to cope with the requirement. Figure VII.5 illustrates the construction of a four-tunnel scove.

The fourth and successive courses of bricks are laid in such a way that rows of brickwork finally meet, and tunnels are thus completed. The progression of the early stages of construction of a scove is shown in figure VII.6. In the foreground, a few courses of fired bricks are set, marking out the tunnel positions. In the middle of the picture, the first corbelled-out course of green bricks is partly set, while further back several courses are laid.

Green bricks are set above tunnel level, in alternate courses of headers and stretchers up to a height of at least 3 m above the ground. At the edge of the scove, each course is stepped in a centimetre or so, to give a sloping side. Small spaces are left between the bricks to allow the hot gases from the fires to rise. The required maximum spacing between bricks is a 'finger width'. This is easy to achieve although a narrower spacing may be satisfactory. As the scove is built up, an outer layer of previously burnt bricks is laid, to provide insulation. This will also allow the proper firing of the outer layers of green bricks.

On the top of the green bricks, two or three courses of previously fired bricks should be laid, bed face down and closely packed. The whole structure should then be scoved with wet mud to seal air gaps. Turves are sometimes laid on top to reduce heat losses. The wet mud should not contain a high fraction of clay if cracks are to be avoided during firing.

Some of the top bricks half-way between the tunnels must not be scoved so that they may be lifted out to increase air flow through the kiln as required. The provision of this adjustable ventilation can be most useful in controlling the rate of burning.

Firewood is set into the tunnels (figure VII.7) for firing. It should preferably be at least 10 cm across, in pieces about 1 m in length. Kindling should be set in the mouth and bottom of the tunnel. Since the heat of the fire is to rise up into the bricks, it is essential that strong winds do not blow through the tunnels, cooling bricks down, and wasting heat. Such winds may increase fuel consumption by 25 per cent. A number of measures may be taken to avoid this waste of heat, including the blocking of the centre of the tunnel during construction, or the temporary blocking of tunnel mouths with

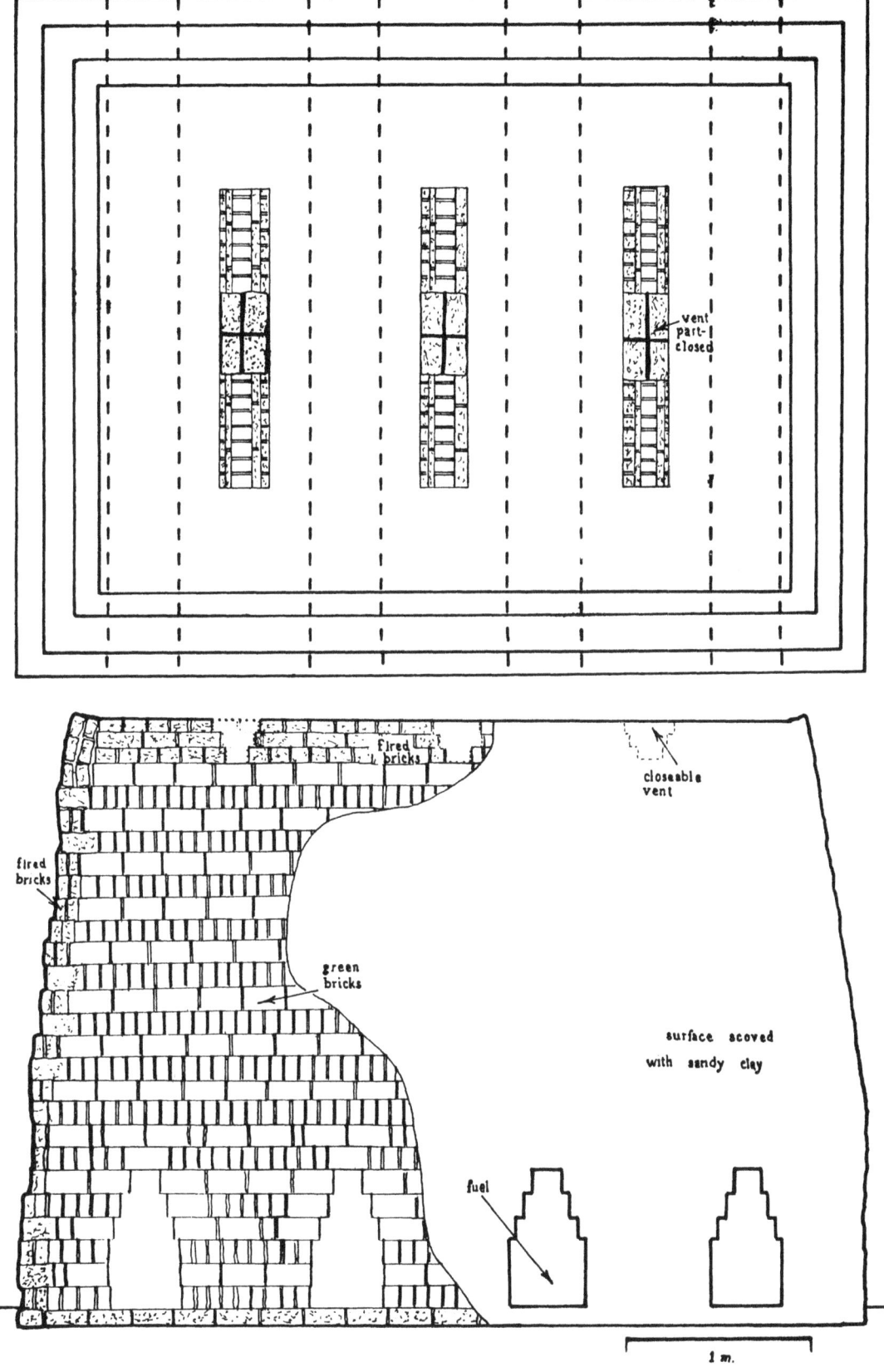

Figure VII.5

Scove: schematic drawing

Figure VII.6

Corbelling scove tunnel (Sudan)

Figure VII.7

Wood in scove tunnel (Sudan)

bricks. In the latter case, one end may be bricked up and fire set at the other end. Once the fire is well alight, that end may be bricked up while the previous one is opened and lighted. Thereafter, the fires may be controlled by bricking up tunnel mouths with loose bricks and adjusting the vents on top. As fuel burns away, it must be replenished.

As with all kilns, heat must be gentle at first until all the water in the bricks is driven off. Adequate air flow is therefore essential to remove the steam produced. Thus, the vents should be open, and the fires kept low so long as steam is seen to rise from the top of the scove. This water smoking period may last several days.

Once the water smoking stage is completed, the fires may be built up gradually to increase temperatures up to a maximum over a period of a few days. A maximum temperature is indicated by the charring of dry grass or paper thrown on top of the scove, or the appearance of a red glow by night. The vents should be closed with fired bricks well before the maximum temperature is attained in order to regulate the burning rate and, thus, help to even out the temperature amongst the bricks. The maintaining of this temperature for several hours (i.e. soaking stage) requires a last charge of fuel, the closure of the tunnel mouths and the sealing of closed vents with mud.

The scove should be left to cool naturally for at least three to four days. Then if necessary some bricks may be removed from the outside to speed the later stages of cooling. Subsequently, the bricks may be left in position until sold. Before collection or despatch, under- and over-fired bricks must be discarded and the remainder, if of variable quality, should be sorted out into good quality, 'seconds' and soft-fired bricks. Rejects may be incorporated in the next scove.

Although wood is generally the fuel used in scoves, oil-burners are used in some countries. Coal, which is also an alternative fuel for scoves, requires a special grate at each end of the tunnel mouth, and is therefore more appropriate for firing in permanent kilns.

The fuel efficiency of scoves is low, 16000 MJ of heat being required per 1,000 bricks for a typical African scove (10, 52). A square scove has a smaller cooling area than a rectangular scove, for a given number of bricks. However, it will require a relatively longer tunnel which may exceed the allowed length for proper lighting of the scove. Thus, small kilns could be square while larger ones may need to be rectangular.

In order to increase the heat efficiency, the height of a scove should be as great as possible, so long as saleable bricks are obtained from the top.

Safety must be borne in mind, however, as high scoves tend to be unstable as a result of shrinkage of bricks during firing. Moreover, a high setting complicates the placing of green bricks on top courses, and increases the risk of accidents. Figure VII.8 shows a high scove of approximately 60,000 bricks, after firing and stripping of the outer bricks.

A scove may be built for firing a few thousand bricks only, but will be less fuel efficient than larger scoves.

III.3 The Scotch kiln

The Scotch kiln is similar to the scove, except that the base, the fire tunnels and the outer walls are permanently built with bricks set in mortar (figure VII.9). The kiln itself has no permanent top, green bricks being set inside the kiln, as shown on the extreme left of figure VII.9. Much basic construction work is thus saved. A plane and section of a seven-tunnel Scotch kiln is shown in figure VII.10. Walls on either side are buttressed, and corners are massively constructed. Access into the kiln is through a doorway in the end walls. This doorway is filled temporarily with closely laid bricks (without mortar) during kiln operations. In some kilns the whole end wall is temporarily erected (40).

Wood is often used for firing the kiln, although oil burners or coal grates may also be installed. The sink of the bricks - after shrinkage - is more easily measured than in the clamp or scove kiln, since the fixed position of the permanent side walls may be used as a reference point. The sink gives an indication of the firing process within the kiln.

The advantages of the Scotch kiln over other permanent kiln structures are its simple design and easy erection. Setting and drawing of bricks are also simple.

The Scotch kilns, like the clamp and scove, are updraught kilns. They have been widely used in developing countries. Their chief failing is the irregular heating and consequent large proportion of under- and over-burnt bricks. This is especially true for clays with a short vitrification range as they cannot be fired without a good temperature control.

Fuel consumption of the order of 16,000 MJ per 1,000 bricks is generally the norm for Scotch kilns (8, 42).

III.4 The down-draught kiln

In the down-draught kiln, hot gases from burning fuel are deflected to the top of the kiln which must have a permanent roof. They then flow down between

Figure VII.8

A high six-tunnel scove (Madagascar)

Figure VII.9

Small Scotch kiln (Madagascar)

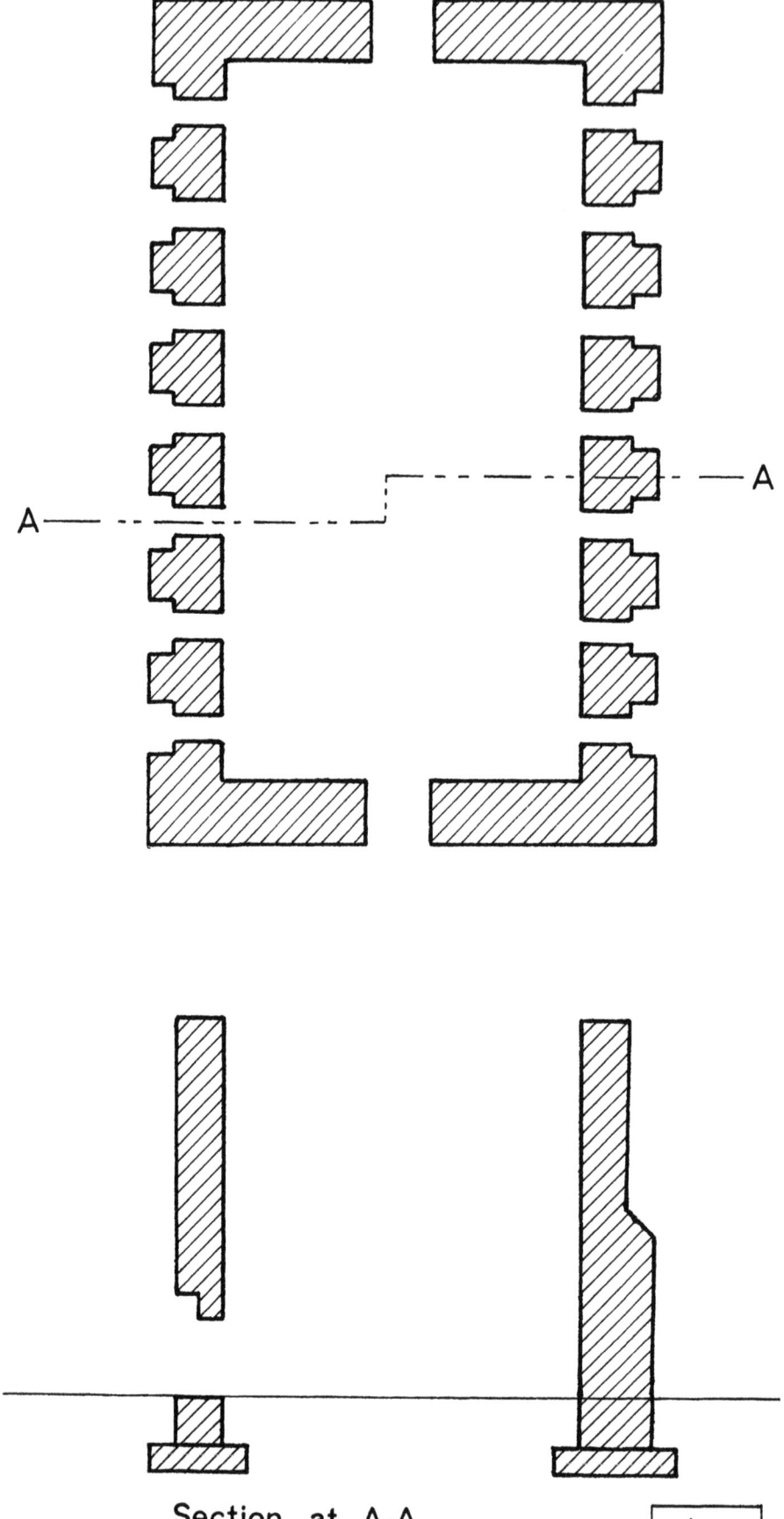

Figure VII.10: Scotch kiln - Schematic drawing

the green bricks to warm and fire them. The green bricks rest either upon an open-work support of previously fired bricks (figure VII.12) or upon a perforated floor through which the warm gases flow. These gases are then exhausted through a chimney outside the area of the kiln after passage through a flue linking the kiln floor to the chimney. The warm gases rising through the height of the chimney provide sufficient draught to pull the hot gases down continually through the stack of green bricks.

The down-draught kiln is more heat efficient than the up-draught kiln described earlier. It can be used for various ceramic products (e.g. drainage pipes and tiles of various types) in addition to the firing of bricks. The kiln can be operated at high temperatures and may then be used for the production of refractory ware.

Circular down-draught kilns may be built in place of rectangular kilns. They are stronger than the latter, but require reinforcement with steel bands to keep the brickwork from deteriorating through periodic cooling and heating. Rectangular down-draught kilns are more simple to build, although they require also steel tie-bars as a reinforcement. They however have the advantage of being easier to set with green bricks than circular kilns. Figure VII.11 shows the ground level plan of a rectangular down-draught kiln of massive construction, with 14 grates for burning fuel. A number of grates stocked with lighted wood are shown in figure VII.13. The grates may be prefabricated from iron bars, as indicated in figure VII.11. A "flash" wall is built behind the grates to keep the flames off the nearby green bricks. The wall in the figure is continuous. Alternatively, separate "bag" walls can be built around the back of each fire (see figure VII.12 right-hand side). The continuous wall tends to even out the heating effect.

Hot gases rise to the arched crown of the kiln and are drawn down between open set bricks (figure VII.12) by the chimney "suction", through the perforated floor (shown in the figure) along its centre line. There should be a few small holes at the base of the flash wall, in the underground flue in order to ensure the burning of bricks near the bottom of the wall. A metal sheet damper is available near the bottom of the chimney in order to vary the flow of gases and exercise control over the operation of the kiln. The control of air flow is achieved by the use of metal doors. These should be thick enough to avoid distortions (figure VII.14).

Entrance to the kiln is through small arched doorways (figure VII.15) referred to as "wickets". These are bricked up temporarily during firing.

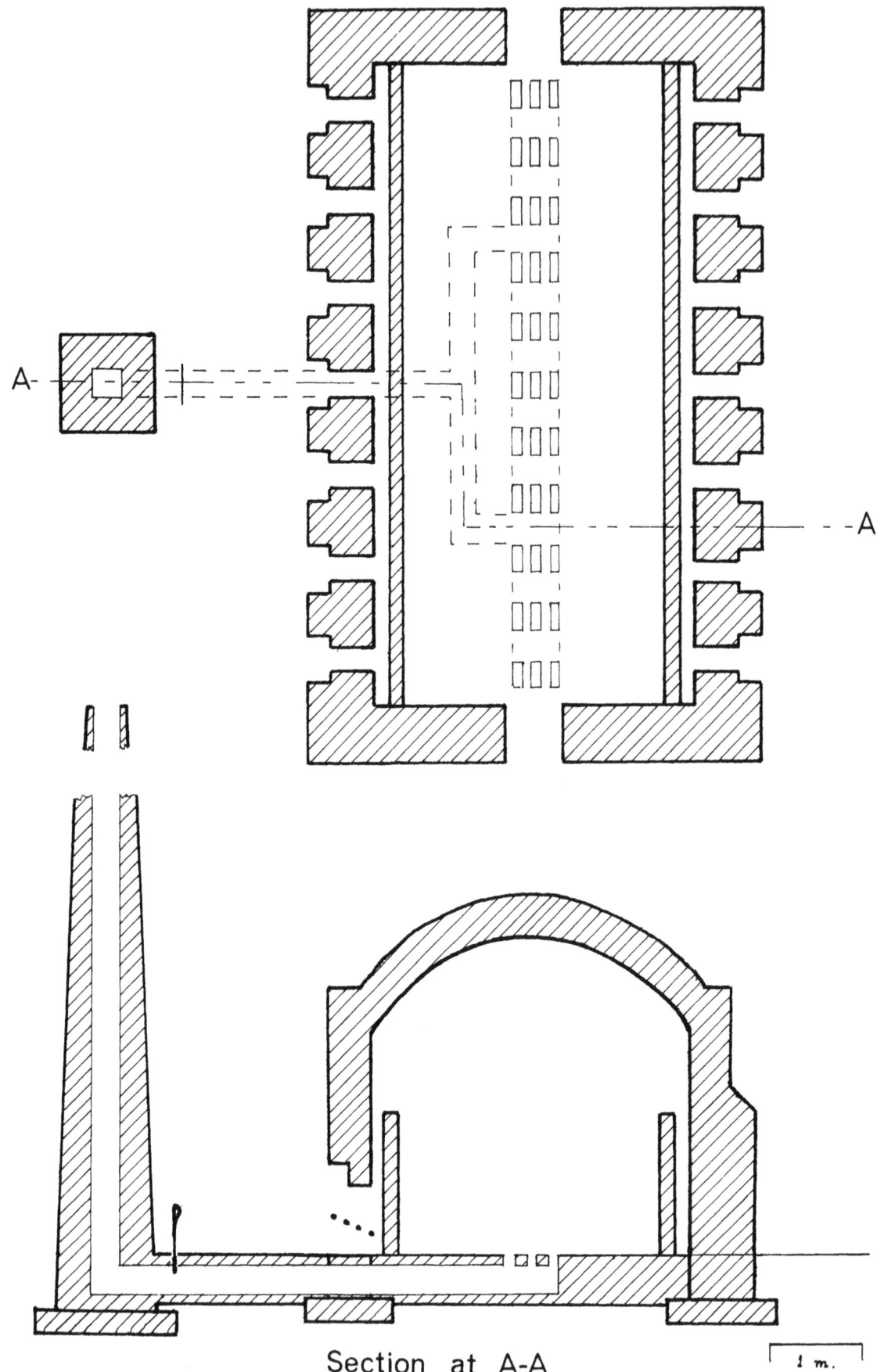

Figure VII.11: Rectangular down-draught kiln

Figure VII.12

Setting and bag walls in down-draught kiln
(Ghana)

Figure VII.13

Fires in kiln grates
(West Africa)

Figure VII.14

Metal damper door for kiln grate (West Africa)

Figure VII.15

Rectangular down-draught kiln with covering (Ghana)

The height of downdraught kilns should not be too great, as it is difficult and time consuming to set bricks at heights that may not be easily reached by workers.

Down-draught kilns may hold from 10,000 to 100,000 bricks. The one shown in figure VII.15 takes 40,000 bricks.

Fuel consumption depends greatly upon the condition of the kiln, the manner of setting the bricks and the control of the firing process. For example, damp foundations absorb heat, and a badly fitting damper may waste fuel. The loss of heat varies from one type of kiln to another. Large kilns consume less fuel than small kilns, for a given number of bricks. An under-filled kiln loses as much heat as one properly filled. An over-filled kiln prevents the passage of hot gases, and this requires a longer burning cycle. Too much draught allows more heat to be wasted up the chimney. Given the above varying circumstances, the heat required by downdraught kilns varies from 12,000 to 19,000 MJ per 1,000 bricks. An exact estimate of heat consumption requires an in-depth study of the characteristics of the kiln (5, 24, 33).

III.5 Original circular Hoffmann kiln

The Hoffmann kiln is a multi-chamber kiln where the air warmed by cooling bricks in some chambers pre-heats the combustion air for the fire, and exhaust gases from combustion pre-heat the green bricks. The main advantage of this kiln is its particularly low fuel consumption rate.

The original Hoffmann kiln was circular (see figure VII.16) and built around a central chimney. An arched-top tunnel surrounds the chimney at a distance of a few metres, and is connected to it by 12 flues passing through the brickwork between the tunnel and the chimney. Each flue can be closed off by dropping a damper. Entrance into the tunnel is through any one of 12 wickets. During operation most of the kiln's tunnel is full of bricks either warming, being fired or cooling.

A typical condition of the kiln is shown in figure VII.16. All but two neighbouring wickets are closed. Cold fired bricks are drawn from one part of the tunnel adjacent to one of the open wickets and dry green bricks are set by the other wicket. Cold air flows to the warm chimney through both wickets. This air cannot pass through the recently set bricks as they are sealed off

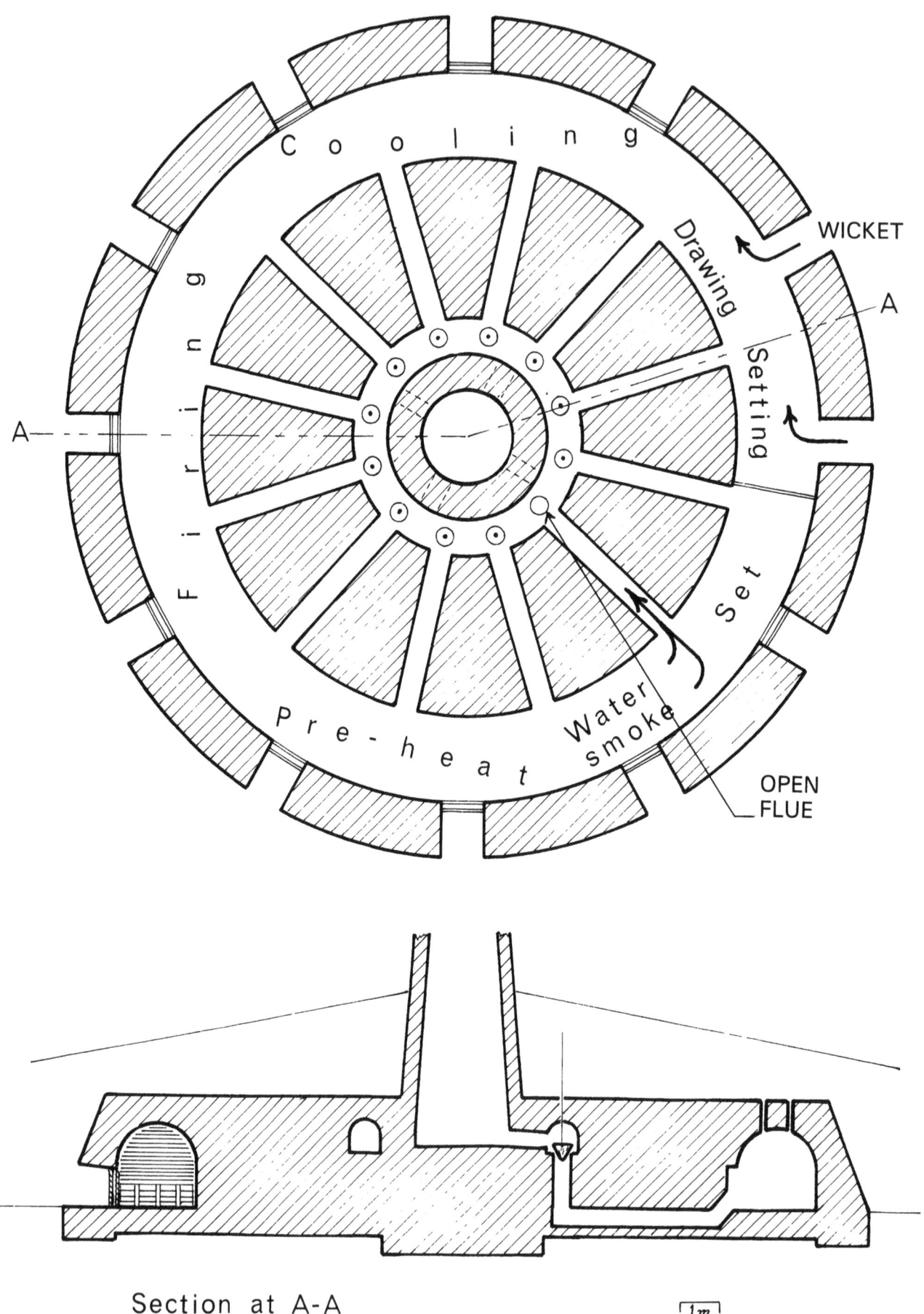

Section at A-A

Figure VII.16

Original Hoffmann kiln

with a paper damper across the whole width of the annular tunnel. The air flows through the bricks which are drawn, into warm bricks further down the kiln (counter-clockwise in the figure) close to the fire. As the air flows counter-clockwise, its temperature rises through contact with increasingly hot bricks. The air is thus pre-heated and ready for efficient combustion in the firing zone of the kiln where fuel is fed in through closeable holes in the tunnel roof. Thus, little fuel is consumed for heating the combustion air. The latter also performs the useful task of cooling bricks for drawing, thus making kiln space available on a relatively short time. The hot products of combustion cannot be vented straight to the chimney through the nearest flue, as the latter is closed (this is indicated by a dot in the circle centre in the figure). Instead, the hot gases pre-heat unburnt bricks. Thus, less fuel is required at the firing stage in order to get the bricks to the maximum temperature. Next, the cooled gases flow through recently set green bricks, bringing the latter to the water smoking stage. These bricks are sufficiently warm to exclude the forming of condensation. Figure VII.16 shows the gases leaving from the open damper (no dot in the circle). Subsequently, this damper is closed, the next one (counter-clockwise) is opened and the bricks marked "set" start the water smoking stage. The fuel feed, and the drawing and setting operations, are also moved counter-clockwise at this stage[1]. Once the part of the tunnel marked "setting" has been filled with green bricks up to the next flue, a paper damper is pasted over the bricks and the wicket (counter-clockwise) is then broken down and cooled fired bricks are withdrawn. The paper dampers can be torn open by reaching through the fueling points with a metal rod.

Figure VII.16 also shows a sectional drawing of the Hoffmann kiln. Bricks in the firing zone are on the left-hand side of the figure, and the empty part of the kiln - between drawing and re-setting - including the closed flue damper, is on the right-hand side. A roof covering protects the kiln from adverse weather. Wickets are shown as two thin walls, separated by an air gap. Thus, heat is kept in the kiln without the need for expensive building work at the wickets.

In the original Hoffmann kiln, fuel fed through the roof falls into hollow pillars formed by bricks set for firing. Ash from the fuel causes some discoloration of the bricks.

---

[1] The Hoffmann kiln described in this memorandum is operated counter-clockwise. Other kilns may, however, be operated in a clockwise fashion.

The tunnel is subdivided into 12 notional chambers which are identified by the flue positions. Each chamber is approximately 3.5 m long and 5 m wide. The height of each chamber is restricted to about 2.5 m for easy working conditions.

Daily rate of production from such a continuous kiln is at least 10,000 bricks.

The advantages of the original Hoffmann design include the identical chambers, the fairly short flues and low fuel consumption (2,000 MJ per 1,000 bricks(8)).

III.6 Modern Hoffmann kilns

Increased demand for bricks in industrialised countries require the erection of substantially larger kilns than those originally designed by Hoffmann. Consequently, the original circular kiln has been modified for the following reasons:
- increases in the floor area of the chambers require considerably more building work between the chambers and the chimney;
- larger diameter kilns and longer flues increased costs considerably and greatly complicated the operation of the kiln;
- the circular shape of the kiln is inconvenient for some sites;
- curved walls make the setting of bricks a difficult operation;
- a circular kiln does not allow the construction of a long tunnel unless the diameter is to be increased considerably. Yet a long tunnel is more appropriate for the transfer of waste heat.

Under these circumstances, the original design was modified into the so-called elliptical Hoffmann kiln, which has long straight walls and a few curved chambers at the end (see figure VII.17). The operating principle is exactly the same as that of the original design. The main difference relates to the larger number of chambers available in the elliptical design.

The operation of the modern elliptical Hofmann kiln may be summarised with reference to figure VII.17. It includes the following sequence of events:[1]

---

[1] The dotted lines in figure VII.17 indicate the boundaries between chambers, and the position of paper cross dampers is shown at inter-chamber boundaries by continuous thick lines.

- 123 -

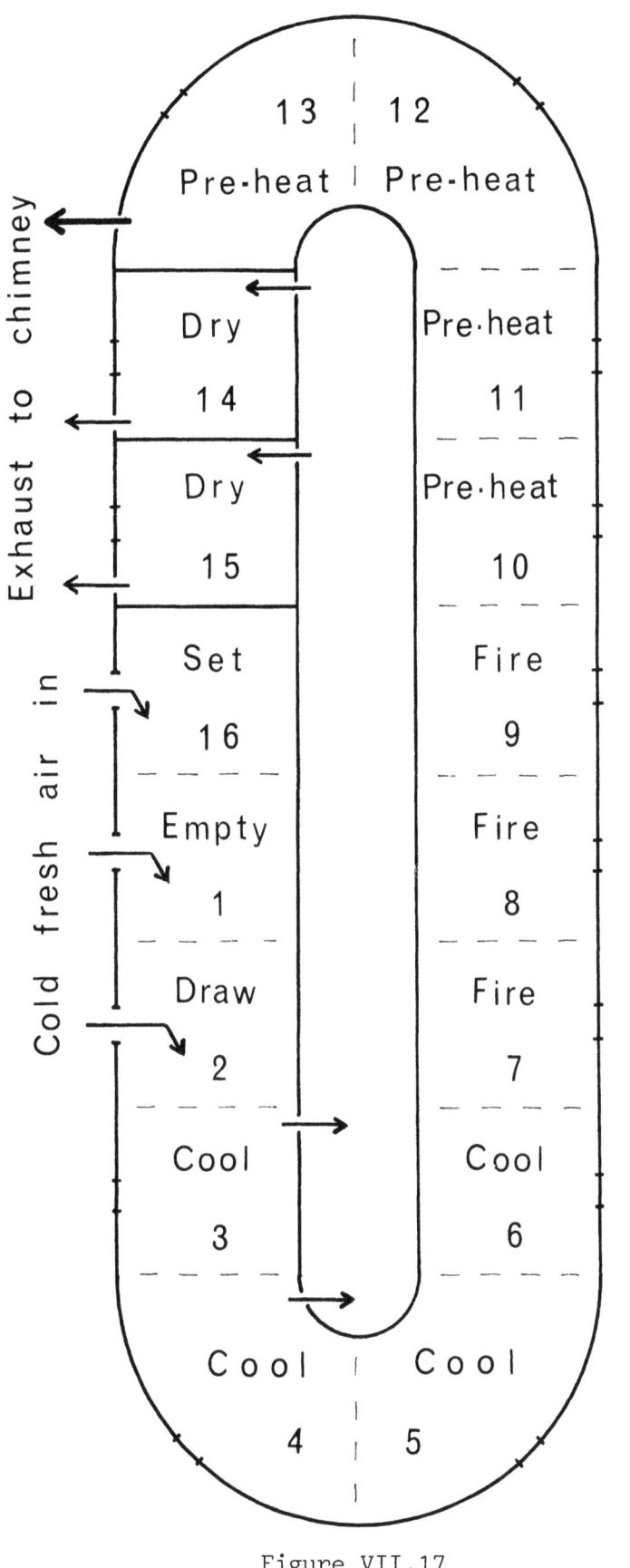

Figure VII.17

Scheme for operating a modern elliptical Hoffmann kiln

- open wickets allow fresh air into chambers 16, 1 and 2 while bricks are drawn from chamber 2;
- other bricks are cooled and air heated in chambers 3 to 6;
- the hot air is used for graduated combustion in the next three chambers 7 to 9;
- exhaust combustion gases are pulled by the action of the chimney through chambers 10 to 13, thus preheating the bricks;
- gases leave the kiln at the end of chamber 13.

In this type of kiln, gases are too cool for water smoking. As they carry much water vapour, there is a risk of spoiling green bricks by condensation (e.g. softening of bricks, surface cracking, and scumming by salts deposited from the products of combustion). Overcoming these problems - which may arise with exhaust gases at less than $120^{\circ}C$ and which are present to some extent in the original circular Hoffmann kiln - requires a second flue which connects all the chambers. Any of the latter may be connected or disconnected to this so-called hot air flue by opening or closing dampers in the same manner as for the main flue connection. In figure VII.17, the hot air flue is regarded as being in the central island of the kiln. Some of the warmed fresh air is taken off by the chimney suction applied to the flue. It is provided by chambers 3 and 4 where bricks are still hot, passed down the hot air flue, then into chambers 14 and 15 where drying, or water smoking takes place. Hence the drying is done with clean warm air, containing no moisture or products of combustion. This air then flows from chambers 14 to 15 into flues (where damper are open) and is exhausted by the chimney. Meanwhile, fresh green bricks are set in chamber 16.

The production rate of most Hoffmann kilns is approximately 25,000 or more per day, a too large output for the type of brickworks considered in this memorandum. However, small kilns can be built to produce only 2,000 bricks per day(8). Figure VII.18 shows hollow clay blocks set within an elliptical Hoffmann kiln with a capacity of about 10,000 ordinary size bricks per day. This is an interesting kiln since wood is used for firing in place of coal (figure VII.19). A wide variety of agricultural wastes may also be used in place of wood in the top-fed kiln shown in figure VII.19. For example, sawdust has been used in Honduras (10).

Fuel consumption of elliptical Hoffmann kilns vary according to the kiln condition and method of operation, as mentioned in the previous section. It is estimated at approximately 5,000 MJ per 1,000 bricks.

Figure VII.18

Blocks in small Hoffmann kiln (Madagascar)

Figure VII.19

Feeding of Hoffmann kiln with wood (Madagascar)

VII.7 Bull's Trench kiln

A large fraction of the cost of construction of the Hoffmann kiln is in the building of the arch of the long tunnel, and in the provision of a chimney, with connecting flues and dampers. Thus, the idea behind the design of an archless kiln by a British engineer (W. Bull) in 1876.

As with the two types of Hoffmann kiln, the Bull's trench kiln may be circular or elliptical. Both forms have been widely used throughout the Indian subcontinent. Construction of this type of kiln is briefly described below.

A trench is dug in a dry soil area which is not subject to flooding. It is approximately 6 m wide and 2 to 2.5 m deep.[1] Alternatively, especially if the soil is not sufficiently dry, the trench may be dug to only half of this depth, while excavated material is piled up on the trench side, and held out off the trench by a brick wall starting at the bottom of the trench (figure VII.20). The total length of the trench is approximately 120 m. It is so constructed as to constitute a continuous trench.

When in operation, the Bull's Trench is full of bricks warming, being fired or cooling. Cooled bricks are drawn and new green bricks are set, while the fire is moved progressively around the kiln. The exhaust gases are drawn off through 16 m high moveable metal chimneys with wide bases, which fit over the openable vent holes set in the brick and ash top of the kiln. These chimneys are guyed with ropes to protect them from strong winds. The type of chimneys shown in figure VII.21 require six men to move them. This figure also shows the method of fueling whereby small shovelfuls of less than 1.5 cm size coals are transferred from storage bins on top of the kiln, and sprinkled in amongst the hot bricks through the removable cast-iron feed holes. Metal sheet dampers are used within the set bricks to control draught.

Figure VII.22 shows the sequence of events diagramatically. The setting of the bricks within the kiln must be such as to allow sufficient air flow between the bricks and wide enough spaces for the insertion and burning of fuel and the accumulation of ashes. However, the whole setting must be sufficiently strong and stable to ensure safe operation of the kiln. The setting in figure VII.20 shows occasional cross link bricks, between the separate bungs or pillars of bricks, tying bungs together. Information on the firing of these kilns is available (54) and kiln designs are standardised(55).

Modifications to this type of kiln have involved the provision of flues from the trench so that chimneys can be moved on rails located on

---

[1] Excavated soil may be suitable for brickmaking.

Figure VII.20

Cross-link bricks between the separate bungs or pillars of bricks in Bull's Trench kiln

Figure VII.21

Bull's Trench kiln: chimney and feeding
(India)

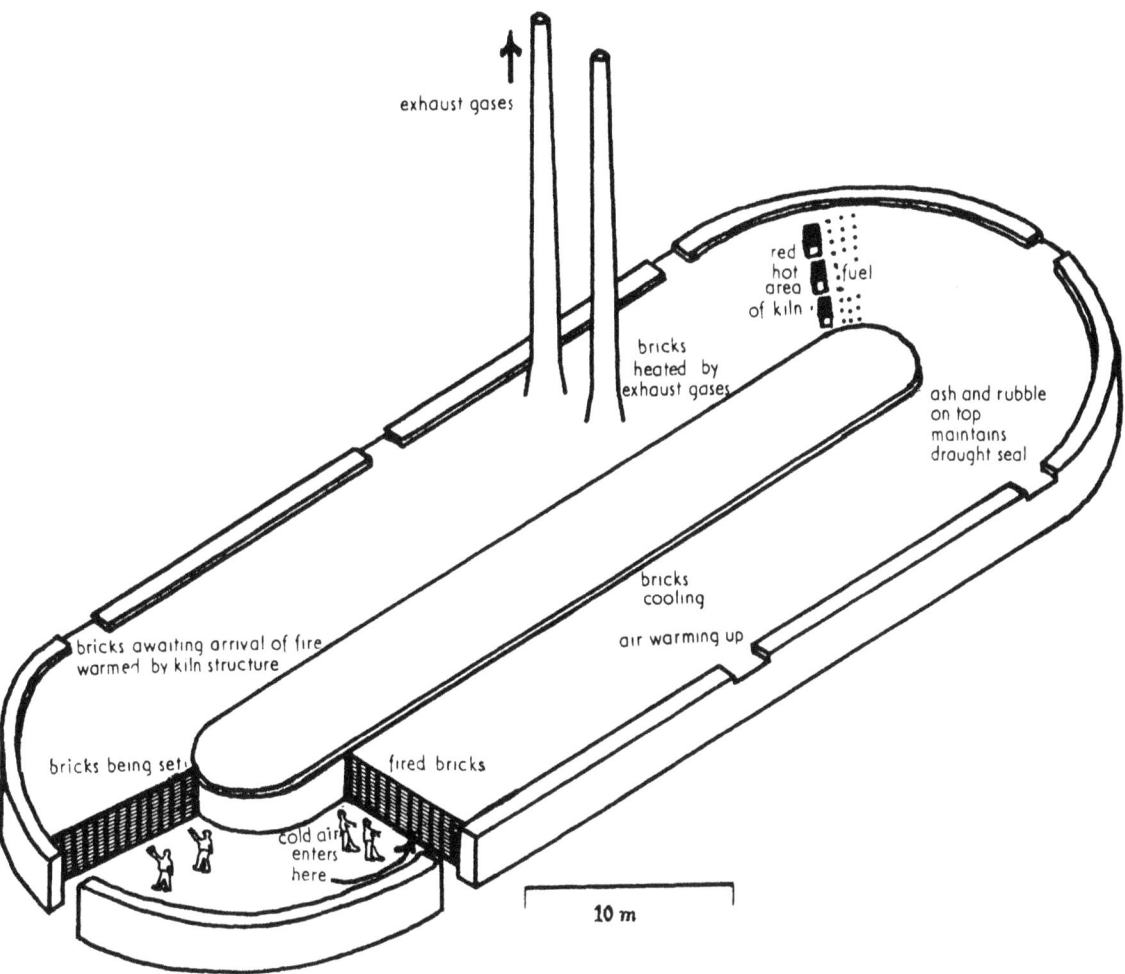

Figure VII.22

Bull's Trench kiln - Firing sequence

the centre island rather than over the setting bricks in the kiln. A major problem is the corrosion of the mild steel chimneys normally used. They may rust through within only a few months. Accordingly, some kilns have been redesigned with dampers opening to flues connected to a permanent brick chimney.

The whole Bull's Trench kiln is very large, a normal output being 28,000 bricks per day. With a narrow trench output could be reduced to 14,000 bricks per day. It is not possible to shorten the trench as this will affect the heat transfer efficiency. The depth of the trench cannot be reduced either without impairing the firing behaviour. The kiln would be very big to roof over, and is most suited to dry weather conditions. The chief advantage of this type of kiln is its low initial construction costs.

Fuel consumption is much better than in intermittent kilns, 4,500 MJ being required for firing 1,000 bricks (10, 22, 56). About 70 per cent first-class bricks can be obtained, the remaining bricks being of poorer quality.

III.8 Habla kiln

The effective tunnel length of the Hoffmann type kiln may be increased by the building of zigzagged chambers. The resulting kilns, known as the zigzag kilns, have a faster firing schedule than the Hoffmann kiln. However, they require a fan - and therefore electrical power - as air must travel a longer path and a simple chimney does not provide sufficient draught for air circulation. Fans provide a more steady draught than chimneys and can be better controlled. They allow a larger transfer of heat to the water-smoking stage, thus saving fuel. However, it is best to avoid condensation on fan blades and subsequent corrosion of the latter by having gases extracted at $120^0 C$. This is especially important if fuel or clay contain sulphur compounds such as pyrites which are transformed into sulphuric acid in the kiln gases.

The zigzag kiln developed by A. Habla is an archless kiln. One additional simplifying feature of this kiln is that the zigzagging walls are temporary structures of green bricks which may be sold after firing. Habla kilns are of various designs: in some kilns the flues are returned from all chambers to the central island while in others, some of the flues are returned to the outer walls. Figure VII.23 illustrates the former type of kiln. For simplicity, it omits the hot air flue which can be carried above the main flue for providing clean drying air.

The Habla kiln is rectangular, but close to a square. The one illustrated in figure VII.23 has chambers numbered 1 to 20, every second chambers being

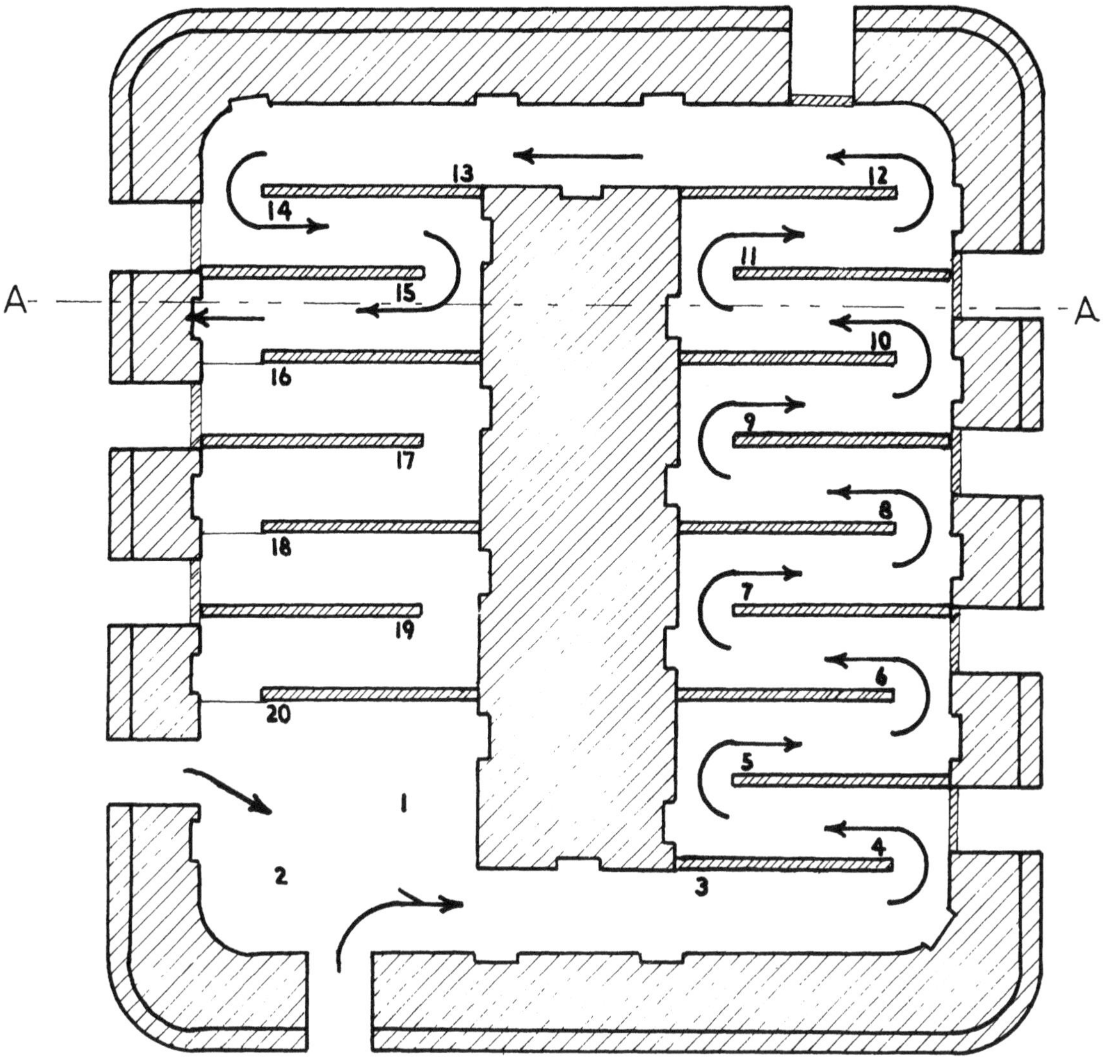

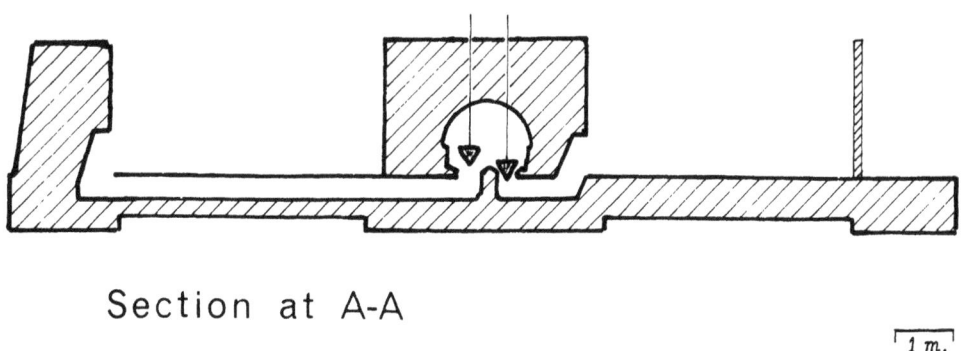

Figure VII.23

Habla type kiln - Firing sequence

accessible through a wicket. Partition walls of dried green bricks, with a thickness of only one brick length, are alternatively built out from the central island and the outer wall. These partition walls deflect the gases from the island to the outer wall, through the wide-set bricks between the partitions. As the temperature of any particular chamber rises, the wide-set bricks are first heated. Then, as bricks in the partition shrink a little, draught through the partitions increases, and the bricks in that partition as well as the wide-set ones become fired.

In figure VII.23, chambers 1 and 2 are empty. Bricks are drawn from chamber 3. Air passing through the open wickets is warmed as it cools bricks in chambers 4 and the following ones. After the firing zone, the exhaust gases preheat bricks, flow out of chamber 15 through the chamber flue and open damper (shown in the section drawing), and enter the main flue from where they are expelled through a short chimney by the fan. Bricks in chambers 16 to 19 are water smoked by clean warm air from chambers 4 to 6. Paper dampers are used as in the case of the Hoffmann kiln.

An interesting modification to the layout shown in figure VII.23 is to build a pair of short partitions in line with each other, approaching from the island and outer wall, but leaving a gap in the middle. Secondly, a wall may be built in the middle of the kiln, separated by two gaps from the island and outer walls. The fire can then travel along two paths simultaneously, around both ends of the second partition, through the central gap of the third, then around both sides of the fourth and so on. This modification should help speed the rate of firing.

Large Habla kilns, producing 25,000 bricks per day (57), have been built in a number of countries. Recently, in India, a 24-chamber high-draught kiln of this type has been developed (58) for an output of 30,000 bricks per day. It is fired with coal and wood. Figure VII.24 is based on published information on this Indian kiln. The latter may be reduced in size for a production of 15,000 bricks per day. Roofed zigzag kilns may also be built for as few as 3,000 bricks per day (8).

The Habla kiln is economical to construct and operate. It has a larger capacity relative to its area than other continuous kilns. This feature reduces the costs of land and construction. Furthermore, the kiln has a long firing zone, allowing difficult clays to be fired more easily. The long-firing path assists heat exchange between gases and bricks, thus improving fuel efficiency. Because partitions are of green bricks, less permanent brickwork has to be heated and cooled, thus adding to fuel efficiency. The shrinkage of bricks in the partitions and consequent leakage of hot gases shortens the distance travelled by the latter. Thus, less power is needed to drive the

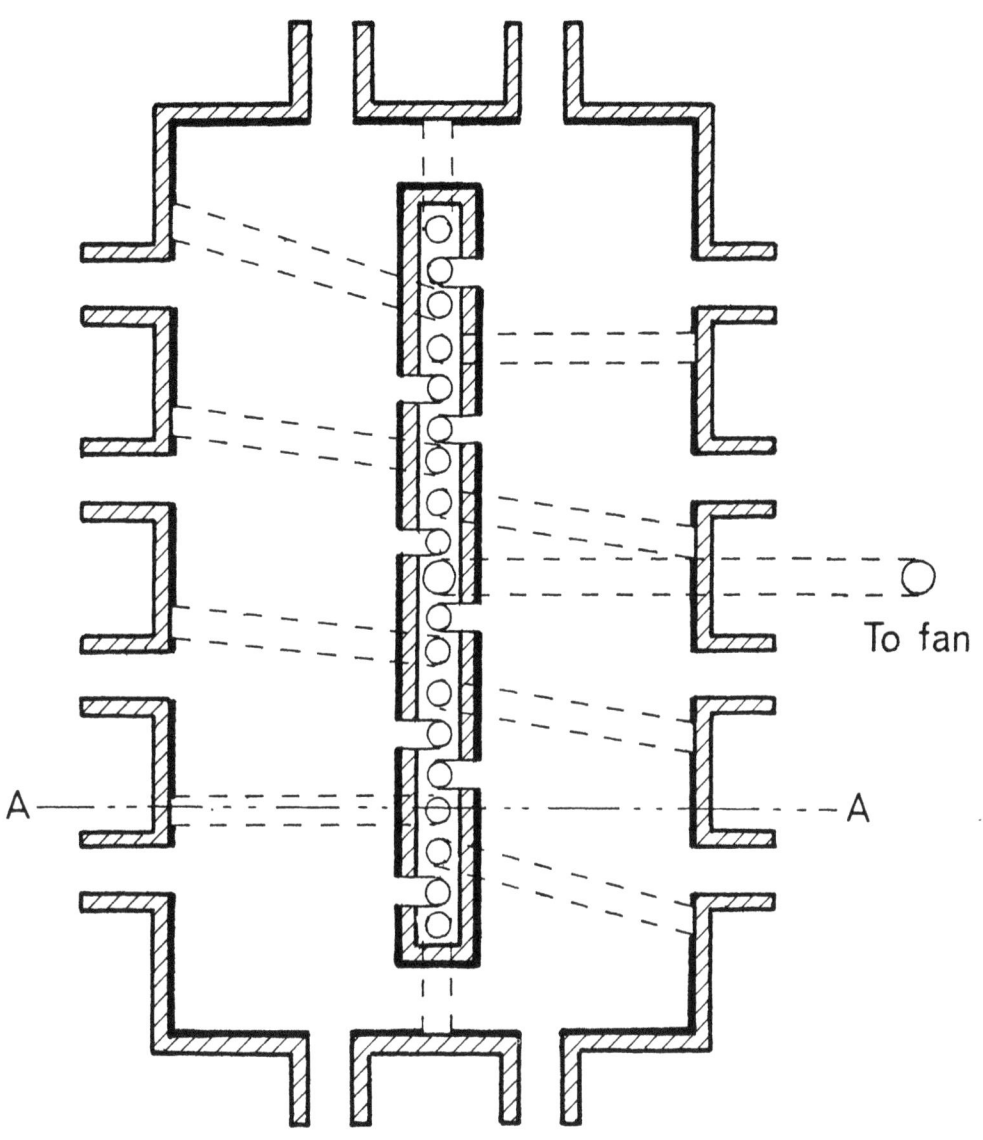

To fan

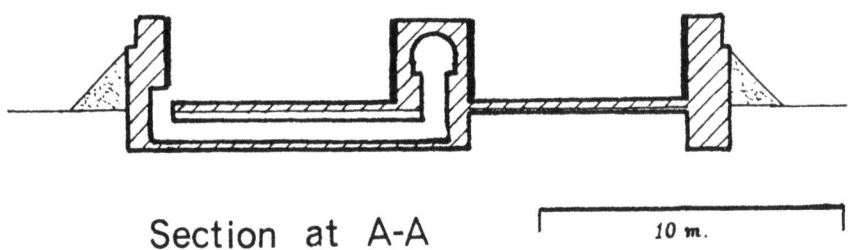

Section at A-A   10 m.

Figure VII.24: Central Building Research Institute high-draught kiln (India)

fan. As a result of partition leakage, the kiln has relatively few "dead spaces" where heat is insufficient to fire bricks properly. The building of partitions of green bricks at the start of each operation does not increase labour costs since the bricks are removed for sale and may thus be regarded as part of the whole setting. Another advantage of the Habla kiln is the easy access to the structure.

Fuel consumption of a zigzag kiln is estimated at 3,000 MJ per 1,000 bricks(57). Consumption in the high-draught Indian archless kiln is also approximately 3,000 MJ per 1,000 bricks.

IV. AUXILIARY EQUIPMENT

IV.1 Kiln control

Temperature control, including control of the temperature attained and of the rates of increase and decrease, is an essential activity in brickmaking. Such control may be carried out in a number of ways, including the following:

Analysis of the colour of gases coming off the kiln: During the initial stage of heating (water-smoking) white gases indicate that the bricks are not thoroughly dry. Thus, the fires must be kept low. The presence of wetness on an iron bar withdrawn from the kiln after only a few seconds' insertion is indicative of the water-smoking stage. From approximately $500^0 C$ upwards, the colour of the hot kiln provides clues to the temperature attained. The interpretation of kiln colours should be based on previous temperature readings and colour examination. For example, the colour of smoke of a scove kiln varies as the whole kiln reaches top temperatures. The sink of the bricks is also an indication of the process of vitrification.

Use of instruments: Instruments are also available to assist the kiln operator. At the water-smoking stage, a thermometer within a protective metal sheath may be pushed in or lowered on a chain amongst the bricks. The presence of moisture on the metal (after withdrawal of the thermometer) is indicative of the water-smoking stage. A reading of the temperature may be obtained by keeping the thermometer amongst the bricks for a longer period. While moisture is still being driven off, the temperature remains at about $100^0 C$. Once it rises to $120^0 C$, fires may be increased as moisture in the bricks will have been removed. Thermocouples may be used in place of thermometers especially if a continuous reading of the temperature is required. A chart recorder, which displays the temperatures measured by several thermocouples, may be used for this purpose. In a continuous kiln, thermocouples could be placed in the preheating zone, the maximum temperature firing zone, in the coolest part of the cooling chamber from which air is being taken to the chimney, in the base of the chimney and in the hot air flue if it exists (48). A reading of the chart will indicate the state of the kiln and the need to adjust firing in order to obtain better quality bricks or to improve fuel efficiency. It also gives the manager an indication of the quality of the kiln operation. Thermocouple probes must be of a corrosion-resistant metal or in protective sheaths.

Pyrometric cones, which better measure the effect of both time and temperature upon clay, may be used in place of thermocouples. They are made from carefully controlled mixes of clays and fluxes, and will therefore deform, or squat at different stages in a heating schedule. For example, a cone which squats at $1,140^\circ C$ with a rise of $60^\circ C$ per hour, will squat at $1,230^\circ C$ with a temperature rise of $300^\circ C$ per hour. Thus, pyrometric cones indicate the way in which fired clay products perform in a kiln. The temperature at which cones squat, when heated at a standard rate, is known as the pyrometric cone equivalent (PCE). Cones may be used in laboratory experiments and in full-size production kilns, and are useful in applying the results of laboratory investigations. In practice, three cones are used to control the firing temperatures within a kiln at a given PCE: one cone of the adopted PCE value, one of $20^\circ C$ less, and one of $20^\circ C$ more. These cones are set in a fireclay base and placed between bricks well in from the kiln wall but in line with a plugged spy hole, sealed with clay. The cones are checked periodically until the first cone, with the lowest PCE rating, begins to bend over. The fire should then be stocked less as the cone with the highest PCE rating should not be allowed to squat. In other words, only the first and second cones should squat. Squatting of the third cone indicates that the kiln has been heated beyond the intended temperature. Sets of cones can be placed amongst the bricks remote from the holes in order to check temperatures attained in any part of the kiln. Alternative systems, using rings or bars, are also used.

Draught gauges, consisting essentially of water-filled, open-ended U-tubes of glass, may be used to measure the draught or suction at any given point of the kiln by connecting one end of the tube to a chamber or chimney and measuring the difference in water levels in the tube. This gives a quick check on the kiln's functioning, and indicates whether dampers should be adjusted.

It is good practice to keep a record book of all kiln operations, including temperatures, draught, numbers and sizes of bricks set, and the number and size of saleable bricks of various grades. The quantity of fuel, operators on duty and any incidental remarks about wind, rain, etc. should be noted in the record book.

IV.2 Brick handling

Labour costs may be substantially lowered if the handling of bricks (e.g. green, dried, and fired bricks) can be rationalised, and if efficient transport devices are used. Furthermore, these devices can be of considerable assistance in relieving the burden of workers. A simple litter borne by two people can be used for the transport of bricks as in the case of clay. The crowding barrow (figure VII.25) is extremely useful, especially in confined spaces such as in the circular downdraught kiln. The barrow is short, with its wheel well placed under the load which may be stacked high. The balance of the barrow is adjusted by placing the bricks according to the workers' height. Thus, little weight is taken on his hands and the load does not tip forward.

V. FUEL

Given the high prices and scarcity of many fuels, achievement of fuel efficiency becomes an important factor in brickmaking. Table VII.1 compares typical requirements of the different kilns described in Section III. It must be appreciated that actual amounts could vary widely from those given, depending upon the size of kiln, size of bricks, nature of clay, firing temperature, the condition of the kiln, and the skill of the operators. Calorific values of wood, coal and oil have been taken as 16,000, 27,000 and 44,000 MJ per tonne respectively. Figures in brackets indicate that the fuel is not suitable for a particular kiln.

Figure VII.25

Crowding barrow

Table VII.1
Typical fuel requirements of kilns

Type of kiln	Heat requirement (MJ/1,000 bricks)	Quantity of fuel required (Tonnes/1,000 bricks)		
		Wood	Coal	Oil
Intermittent				
Clamp	7,000	(0.44)	0.26	(0.16)
Scove	16,000	1.00	0.59	0.36
Scotch	16,000	1.00	0.59	0.36
Downdraught	15,500	0.97	0.57	(0.35)
Continuous				
Original Hoffmann	2,000	0.13	0.07	0.05
Modern Hoffmann	5,000	0.31	0.19	0.11
Bull's Trench	4,500	0.28	0.17	(0.10)
Habla	3,000	0.19	0.11	(0.07)
Tunnel	4,000	(0.25)	(0.15)	0.09

Source: 5, 8, 10, 22, 24, 33, 44, 52, 56, 57, 58.

Where wood is used as a fuel, trees should be replanted to replenish supplies. Sometimes, commercial ventures provide an incidental supply of wood (e.g. in the rubber estates, where trees fulfil their latex-yielding life within 30 years or so). The practice of coppicing is worthwhile: small-size wood is cut from a low level, thus allowing the trees to continue to grow. Coppicing may yield 125 tonnes of wood every year from a square kilometre (22).

Many other materials can be utilised to assist in the firing of bricks, in addition to wood, coal and oil. Gas, either naturally occurring such as Sui gas in Pakistan, or made from plant wastes by gasification or by bio degradation processes (biogas), may be burnt in simple burners, preferably set low down in the kiln.

Agricultural and plant wastes may also be used for firing, including sawdust (e.g. in Honduras (10)) and rice husk (e.g. in experimental kilns and in a Bull's Trench kiln in Pakistan). A large quantity of ash, some 18 to 22 per cent of the weight of the husk, is produced on burning. As this ash may block the bottom of the Bull's Trench kiln, coal is used without husk on every few blades. Groundnut husks, coffee husks, chaff, straw and coconut husks have also been used on an experimental basis. They may be burnt in the grate of an intermittent kiln, or fed in through the top of Hoffmann, Bull's or Habla kilns. However, the volume required for sufficient heating is very large in most cases. These wastes may not be easily burnt in kilns equipped with grates. They may therefore constitute only part of the fuel source. Dung is a traditional fuel in Sudan.

Waste materials may also be mixed with clay (e.g. 5 to 10 per cent by weight) instead of being directly used as a fuel source. While this method is technically feasible, it may produce more porous and weaker bricks.

Other wastes have been tried in a number of kilns. These include old rubber tyres, waste engine oil, ashes, clinker, pulverised fuel ash from coal-fired power stations, and coal washery wastes.

The optimum air flow for highest fuel efficiency in a kiln can be determined by experiment, especially if good records are kept. Adequate air is necessary to obtain full combustion of fuel. However, too much air will have a detrimental cooling effect.

The best use of fuel is generally obtained in continuous kilns, properly maintained and run, using a fraction of waste materials. Economy of scale favours the larger kilns, but fuel for transportation of bricks from a large-scale plant may negate the fuel savings from the operation of the kilns.

VI. PRODUCTIVITY

The range of outputs of the various kilns described in Section III is shown in table VII.2 Although smaller units might be built than those quoted, lack of data on these units precluded their inclusion in the table.

Table VII.2
Range of outputs from various kiln types

Type of kiln	Capacity (bricks, '000s)	Capacity (bricks per day,'000s)
Intermittent		
Clamp	10 - 1,000	
Scove	5 - 100	
Scotch	15 - 25	
Downdraught	10 - 50	
Continuous		
Original Hoffmann		10 - 15
Modern Hoffmann		2 - 24
Bull's Trench		14 - 28
Habla		15 - 30

Many factors affect the obtained outputs. In general, good maintenance and provision of roofs over kilns will improve the quantity of bricks produced, but will add to costs. Quantities produced will also depend upon the required quality standard and the skill of the operators. The market demand, weather, infrastructure and supply of raw materials will govern the rate at which production may proceed.

Capital-intensive equipment is available for transportation and the setting of bricks, but all the kilns described in Section III may be operated on a labour-intensive basis. All but the downdraught and Hoffmann kilns involve the extra task of covering the green bricks with previously burnt ones and ashes.

Most of the kilns described in this memorandum are operated on a fairly labour-intensive basis. Table VII.3 shows the difference in labour requirements between these kilns and more automated kilns used in industrialised countries.

Table VII.3

Labour requirements

Type	Location	Labour requirements for firing, including drying (man-hours/1,000 bricks)
Traditional plant with clamp	Lesotho	15
Traditional plant with coal-fired clamp	Turkey	16
Moderately mechanised plant	United Kingdom	1.5
Highly automated plant	United States	0.6

Source: 10

VII. BRICK TESTING

VII.1 Purpose of testing

The purpose of testing is to check the production process; to remedy faults to ensure a saleable product; and to guarantee the quality and performance of marketed bricks. Given the intended use of bricks in housing construction, they must resist local weather conditions and should not contain materials which will damage them or the applied finishes such as renderings, plaster or paint. They must be strong enough to withstand both the dead load of the building itself, and the live load imposed by occupancy or wind. The thermal or moisture movements should not also be so large as to build up unduly large stresses. Testing should therefore consist of ensuring that bricks have all the required characteristics for efficient use in building.

VII.2 Initial checks on quality

A quick check of quality consists in striking two hand-held bricks. A high-noted ring indicates that they have been thoroughly fired. On the other hand, a dull sound indicates either cracked or soft-fired material. Similarly, bricks which cannot be scratched and rubbed away with the edge of a coin are hard and of good quality. Nevertheless, bricks without a good ring or which can be marked may still be perfectly suitable for many construction purposes. These two tests only serve to identify some of the best materials.

The general appearance of bricks may give a quick indication of quality. Regular shapes and sizes, sharp arrises, unblemished surfaces and freedom from cracks are signs of good bricks. Colour is difficult to interpret. However, within a batch of bricks from any one brickworks, the darker colours are likely to relate to the harder fired, stronger and more durable bricks.

VII.3 Standard specifications

Many countries have published their own standard specifications for bricks, and refrerence should be made to these where available. The methods of testing described in the British Standard(37) have been devised after much research and many years of experience, and are of wide interest. However, they may not necessarily be appropriate for other countries. Actual numerical limits must be decided locally according to what is required and what can be made.

VII.4 Sampling

The testing of bricks should be made on a representative sample of the latter since variations in preparation of raw materials, drying and firing produce bricks of variable characteristics. The testing sample should preferably contain 40 bricks picked up at random while they are being unloaded from the kiln. It is more difficult to obtain random samples once they have been placed in a large stack.

VII.5 Dimensions

Since bricks are used in fairly long runs, the testing of brick dimensions need not be carried out on individual bricks. Instead, 24 bricks should be chosen at random from the sample 40 bricks and small blisters or bumps knocked off. The 24 bricks should then be placed against an end stop touching end to end on a long bench. If the nominal length of each brick must be, for example, 21.5 cm, the far end of the row of 24 bricks should be 516 cm from the end stop. The permissible variations (for example 508.5 to 523.5 cm as in the British Standard for 21.5 cm bricks) should be clearly marked above the level of the bricks on a board behind the bench. Similarly, the width and height of the 24 bricks should be checked as bricks which comply with the standards for one dimension do not necessarily comply in other dimensions. Table VII.4 gives the dimensions calculated from batches of African bricks and the requirements of a relevant Standard: none complies in all three dimensions.

Table VII.4
Overall dimensions of 24 bricks

Brick source	Length (cm)	Width (cm)	Height (cm)
1	530.6	250.2	181.1
2	539.0*	253.2	184.2
3	542.3*	265.2	177.8*
4	543.8*	255.5*	168.4
Standard			
Minimum	533.4	254.0	170.2
Maximum	548.6	264.0	180.3

*Dimension complies
Source: 51

VII.6 Compressive strength

The compressive strength of harder fired bricks is greater than that of other bricks although the strength of most bricks is likely to be adequate for simple buildings. Even a fairly weak brick with a compressive strength of only 7 MN/m^2 may support a 300 m column of bricks without crushing. However, loadings in actual buildings are increased in supporting pillars and around wall openings.

As the strength of dry bricks is higher than that of wet bricks, the former should be immersed in water for 24 hours before testing. Although a simple wooden beam, used as a lever, can be sufficient to crush green bricks or adobe, it is unlikely to be successful for higher strength burnt bricks. Instead, a compression testing machine is commonly used in public works departments, universities, and technical and research institutes. To avoid high local loadings due to small irregularities on the bed faces of bricks in contact with the steel plattens of the machine, thin plywood plattens should be placed on both bed faces. Bricks with frogs or large perforations should be bedded in mortar which is allowed to set prior to testing. Ten bricks should be tested, and the mean strength calculated.

Table VII.5 gives test results obtained on batches of bricks from two different works in East Africa. They were tested by the British Standard method. Although a few individual bricks from one works are weaker than required for a particular grading (not the lowest), the mean value complies with the standard requirement. Similarly, a few bricks from the other works would not meet a requirement of the Indian Standard(36), but the mean value is satisfactory.

Table VII.5
Compressive strength of 10 bricks

Brick source	Compressive strength MN/m^2			
	Individual bricks	Mean	Minimum requirement British Ordinary	Indian Class II
A	4.1, 4.0, 8.1, 8.1, 11.1, 10.9, 7.7, 4.0, 6.1, 7.1	7.1	5.2	7.5
B	6.2, 6.5, 6.5, 8.8, 11.4, 7.7, 11.5, 6.3, 10.3, 9.9	8.5		

Source: 61

For calculated load-bearing structures and civil engineering works, the strength of the bricks should be determined and used in the design calculations.

For non-load-bearing partitions, bricks of only 1.4 MN/m^2 may be used(37).

A low-cost impact testing method, claimed to give useful information on brick strength, consists in dropping a weight several times on the pieces of a broken brick in a containing cylinder(62).

The density of dry bricks is easily determined, and is sometimes quoted instead of strength.

VII.7 **Resistance to erosion by water**

Although compressive strengths may be more than adequate, dampness and flowing water can cause severe deterioration in buildings. Unfired earth bricks or adobes are eroded or slaked by water. On the other hand, firing yields bricks with excellent resistance to water erosion. The testing of resistance to erosion by water requires the soaking of brick samples in water for 24 hours. Good bricks should show no sign of softening or slaking.

A more severe test consists in spraying water on bricks for several weeks. This test allows the separation of bricks into various quality groups as low-quality bricks get eroded while high-quality ones remain unaffected.

VII.8. **Water absorption**

Hard fired bricks will absorb less water than other types of bricks. The quantity of water absorbed by a dry sample of bricks, when immersed in water for 24 hours or eight days(7), or boiled for five hours(37) is often estimated in testing laboratories. However, the results of this simple test are not easy to interpret. Generally, absorptions of less than 15 per cent by weight in the cold tests would be indicative of satisfactory brick strength and durability. Exceptionally, higher absorptions may be found with some types of clay although the bricks may still be satisfactory.

VII.9 **Rain penetration**

In practice, moderately high water absorption is acceptable in a brick wall, since rain water dries out once good weather conditions return. On the other hand, under some circumstances, rain running on walls made of low absorption bricks may enter the wall surface through small cracks between bricks and mortar. Rain penetration tests may be set up by building walls exposed to either rain or a water spray on one side, and observing the other sheltered side of the wall. The quality of not only the bricks but also of the workmanship in laying them, is of significance in determining the performance of the brickwork.

VII.10 Efflorescence, soluble salts and sulphate attack

Appearance of efflorescence on brickwork is considered unsightly and in extreme cases may cause spalling of faces of bricks. It may occur on initial drying or after subsequent wetting of the bricks. One test for determining the presence of efflorescence consists in half immersing bricks in distilled water for two weeks. Soluble salts, if present, will dry out on the top corners of the bricks as water is being soaked up. If only slight efflorescence occurs, it may be assumed that no problems are likely to arise in practice(7). Alternatively, sample bricks can be covered with a polythene sheet to prevent evaporation from the non-visible face in the finished brickwork, while exposing the other face upwards. A bottle of distilled water is then inverted on this latter face and the water allowed to soak in. Subsequently, the water dries out, and salts present in the bricks are carried to the surface where the amount of such salts may be estimated. Generally, if more than half of the exposed area is covered with salts, or if the surface flakes, the bricks would be regarded as efflorescent(37).

The nature of soluble salts can be determined in the laboratory by standard methods of chemical analysis carried out on powder obtained by drilling or fine grinding of brick samples. More than 3 per cent by weight of salts is considered a high concentration. For special quality brickwork, acid soluble sulphates should not exceed 0.5 per cent, calcium 0.3 per cent, magnesium 0.03 per cent, potassium 0.03 per cent and sodium 0.03 per cent(37).

VII.11 Lime blowing

Hydration of quicklime particles derived from limestone in brickmaking clays can cause pitting on brick faces. When, in spite of precautions in manufacture, a problem persists, it may be possible to alleviate it by docking the bricks (i.e. by immersing them in water(50)). Docking apparently slakes the lime to a softer form which may extrude into neighbouring brick pores. Bricks should be soaked for approximately 10 minutes so that water penetrates at least 15 mm. Otherwise, the problem may worsen. Large quantities of water are required, and the increased weight of the bricks may add to transportation costs. Efflorescence may appear as the water dries out.

Testing for lime blowing can be done by immersing brick samples into boiling water for 3 minutes(50), or preferably into a steamy oven(63).

VII.12 Frost

Frost is one of the most destructive natural agents but only when brickwork is frozen while in a very wet condition. In climates where frost occurs, outdoor test walls or laboratory freezing and thawing tests can be used to assess frost resistance(10, 64). Harder fired, less porous bricks are generally more resistant to frost.

VII.13 Moisture and thermal movement

Reversible and irreversible moisture movements are likely to be small and of little consequence for small buildings. Tests can be carried out to determine these movements if bricks are to be used for long runs of brickwork, or in tall structures. Thermal expansion is of similar significance and may be measured: it is likely to be approximately 4×10^{-6} m per $^{\circ}C$ (i.e. a change in temperature of $20^{\circ}C$ will cause a 12 m run of brickwork to change 1 mm in length).

VII.14 Durability and abrasion resistance

To investigate performance, including resistance to abrasion by wind-blown sand, small test walls should be constructed outdoors from the various types of bricks made at any works. Changes in the brick surface over various periods of time will indicate the degree of resistance of bricks to abrasion.

VII.15 Use of substandard bricks

Bricks that do not meet the required standards may be used for other purposes, and not be entirely wasted.

Underburnt bricks may be returned for refiring, or may be useful in kiln construction. Overburnt bricks may also be used in kiln construction or can be broken up and used as concrete aggregate.

Reject bricks may be broken up and used for road building or soakaways. If finely ground, they may be used as grogs in brickmaking.

Reject, underfired bricks exhibit pozzolanic properties when ground down to powder. The latter may thus be mixed with lime for the production of a cement substitute. Alternatively, the lime released when ordinary Portland cement sets will react with the brick powder. Thus, crushed soft-fired clay could constitute a useful mortar ingredient.

CHAPTER VIII

MORTARS AND RENDERINGS

I. <u>PURPOSE AND PRINCIPLES</u>

Mortar is used to accommodate slight irregularities in size, shape and surface finish of bricks, thus providing accuracy and stability to a wall. In so doing, gaps between bricks are also closed, thus excluding wind and rain and increasing the wall strength.

Rendering applied to the external surface of brick walls can help prevent ingress of rainwater into a building. However, if materials are of good quality and bricklaying techniques are correct, rendering is usually unnecessary. Rendering is preferred, in some countries, for mainly aesthetic reasons while bare fair-faced brickwork is preferred in others.

In general, mortars need not be stronger than the bricks, and renderings should be mixed and used so that they do not crack (see section III).

Mixes for mortar and render are frequently made from ordinary Portland cement or lime, together with a large proportion of sand. In principle, a good solid mortar should contain two-thirds sand and one-third binder. If less than a third is used, the wet mix will be less workable and less strong when set. If more than a third is used, greater shrinkages will occur, and the risk of cracking will increase. Furthermore, the cost of mortar will be unnecessarily high.

II. MORTAR TYPES

II.1 Mud

The most elementary mortar, mud, is made from soil mixed with water. It may be suitable for laying adobe, but is not recommended for fired bricks. Mud mortar exposed to the weather in fair-faced work will quickly be eroded by wind-blown sand and rain (figure VIII.1). A good-quality rendering is essential if mud mortar is used. However, cement used in the render mix will be better utilised in making a more durable mortar.

II.2 Bitumen/mud

The addition of bitumen as a cut-back or emulsion makes a mud more water-resistant. Asphalt may be used as an alternative material.

II.3 Animal dung

Mud renderings may be made more weather-resistant by the incorporation of cow dung. A thin paste made by adding water to a mix of one part cow dung with five parts of soil may also be used to wash over a mud rendering(20).

II.4 Lime/sand

Lime and sand mixes are traditional materials. Lime varies in purity and thus gives different types of mortar. If the lime is very pure, consisting of a large proportion of calcium hydroxide, the hardening of the mortar will be due solely to carbonation, caused by the slow reaction with carbon dioxide in the air. On the other hand, impure limes often contain a proportion of siliceous material from clay contained in the limestone. In this case, lime burning yields a hydraulic lime, which sets under water if needed. The hardening is in this case a reaction between silica and calcium hydroxide which gives calcium silicate. Hydraulic limes also carbonate in air. This type of mortar can be very good, but slow hardening makes it less attractive than cement mortars. Replacement of some lime by cement gives a useful increase in early strength.

It is essential that lime be completely slaked before use. Commonly, slaked lime may be purchased as such, and quality may well be satisfactory. Alternatively, quicklime can be used, but it must first be mixed with water in a pit. A slight excess of water should be added, and the mixture covered to

Figure VIII.1

Erosion of mud mortar after two years
(West Sudan)

prevent drying out. Several days in the slacking pit are needed for its hydration. If any particles of unslaked material remain, they may slake later in the set mortars, and spoil the mortar.

II.5 Pozzolime

In the same way as silica from the clay heated during lime-burning could react with lime as described earlier, so naturally occurring volcanic ashes may contain siliceous material which can have a pozzolanic reaction with lime. In Tanzania, three parts of ash and one of lime are mixed together to form a "cement". The latter is then mixed with sand for the production of mortar(65).

II.6 Rice husk ash cement

Rice husks burnt at temperatures below $750^{\circ}C$ yield approximately 20 per cent of their weight in pure ash. A cement-like material can be produced by mixing two parts by weight of this ash with one part of lime (66). One part of this rice husk ash cement may then be mixed with three parts of sand (by volume) and water for the production of mortar or rendering.

A similar material can be made from rice husks and lime sludge waste derived from sugar or paper industries(67).

II.7 Brick-dust/lime

Brickmaking clays fired to only $700^{\circ}C$ produce a pozzolanic material when crushed to a fine powder. Two parts of the latter (known as surkhi in India) are mixed with one part of lime for the production of a cement-like material. The latter may then be mixed with sand for the production of mortar (68).

II.8 Ordinary Portland Cement(OPC)

OPC is widely used for mortars and renderings. Excessively strong mixes may be harmful and unnecessarily expensive. Where sulphates from bricks cause problems by reacting with OPC mortars, the use of sulphate-resisting cement might be considered.

II.9 Pulverised fuel ash

Pulverised fuel ash from modern coal-fired electricity generating plants exhibits pozzolanic properties. 30 per cent of this material may be mixed with 70 per cent of ordinary Portland cement for the production of pozzolanic cement. Pulverised fuel ash may also be used with lime. The Indian standard(69) for this and other pozzolanic materials requires that mortar cubes of one part pozzolanic mixture and three parts sand by weight show an initial set in not less than two hours, and a final set within one, one-and-a-half, or two days, depending upon grading. For these three grades, 28-days compressive strengths should be at least 4, 2 and $0.7 MN/m^2$ respectively.

II.10 Plasticisers

Mixes of ordinary Portland cement and sand are made more workable by substituting lime for some of the cement. The wet mix is then buttery and easy to spread by the masons. Instead of lime, very small additions of purpose-made vinsol resins may be added. The latter form minute bubbles during mixing. However, it is difficult to obtain a mortar with the required properties as the mixing operation cannot be easily controlled. Factors affecting the properties of the mortar include the amount and hardness of water in the mix; cement content; fineness and composition; grading of the sand; and the efficiency and duration of mixing. Unless all these factors are properly controlled, problems may arise. For example, the mortar may squeeze out after several courses have been laid, and the brickwork may get out of vertical alignment if too much air is incorporated in the mix. Furthermore, strength may decrease and mortar may become too permeable to water after setting. On the other hand, too little air will reduce the mortar workability. Thus, spreading of mortar may not be carried out properly, resulting in low strength brickwork and poor resistance to rain penetration(70).

Masonry cements (made with ordinary Portland cement and plasticisers) should show an initial set in not less than 45 minutes, and a final set within 10 hours, if they are to comply with the British Standard(71). A 28-days compressive strength should not be less than $6 MN/m^2$.

II.11 Gypsum plaster

Any substance which sets from a fluid into a solid may be regarded as cementitious(72). Gypsum plasters, made by heating naturally occurring (or industrially produced) gypsum will set quickly. They have been used as mortar in arch building without centring. Being slightly soluble in water they are not suitable for exterior use in wet climates. They are widely used for interior wall finishes.

III. MIXING AND USE

Dry ingredients of mixes should be measured out carefully. Although weighing may be preferred, gauge boxes are often used to obtain constant proportions by volume. However, if the water content of sand varies, gauge boxes may not provide accurate mixes.

The dry ingredients should be first mixed thoroughly prior to final mixing with water. Mixing may be done by hand with spades, or in a mortar-mixing machine.

If mortars are made very strong, which is usually accomplished by using a high proportion of ordinary Portland cement, the resulting brickwork may be less able to accommodate any slight movement. For example, edges of bricks may be damaged and any cracking will probably go through the bricks themselves if movement takes place. A weaker mortar might yield a little more under stress, and any cracking would then preferably go through the mortar joints. Such cracking is easier to repair in fair-faced work. For high strength work both bricks and mortar will need to be strong. Table VIII.1 provides the properties of various mixes of ordinary Portland cement (OPC), lime and sand.

Table VIII.1

Mortar mixes

Mix proportions by volume					Compressive strength 28 days (MN/m^2)	Ability to accomodate movement
OPC	Lime	Sand	With plasticiser (small amounts) OPC	Sand		
1	0-.25	3	-	-	11.0	Least able
1	.5	4-4.5	1	3-4	4.5	
1	1	5-6	1	5-6	2.5	
1	2	8-9	1	7-8	1.0	Most able

Source: 73

Strong renderings are more likely to shrink and crack than weaker ones. Cracks in rendering (figure VIII.2) allow water to get into the brickwork, which may not dry out easily. A mix of one part cement, one part lime and six parts sand by volume is excellent for rendering purposes. A 1:2:9 mix may also prove satisfactory. Water content should be kept as low as possible, since wetter mixes shrink more on drying, thus increasing the risk of cracking.

The mix should be used while still fresh, especially if based on OPC. A good mortar will hang on the mason's trowel, then spread easily on the bricks. It may be necessary to kill the suction of the bricks by dipping or splashing them with water, thus preventing a large proportion of the mixing water being instantly pulled out of the mix as soon as it touches the bricks. If much water is lost, it will not be possible to spread the mix as either mortar or render, and there may also be insufficient water in the mix to allow the hydration reactions to take place properly. For the same reason, it would be best to avoid working in the full sun, and to keep the work damp for 24 hours to allow curing to take place. On the other hand, if the mortar is too wet, it may have higher porosity, greater shrinkage, and lower strength, and the appearance of finished work may be poor.

Wide mortar joints are sometimes used in brickworks (figure VIII.3). If bricks are badly shaped, this is unavoidable. Where possible, joints should not be wider than 1 cm if one is to economise on materials and labour. Renderings may be of one or two coats, depending upon the required quality of surface finish.

In the building of brick walls, frogged bricks may be laid frog down if the use of mortar is to be minimised. On the other hand, maximum strength is achieved with bricks laid frog up. If the mortar bed is furrowed before bricks are set into it, the strength of brickwork will be reduced(74). Vertical joints (or perpends) between bricks should be completely filled with mortar to obtain the best resistance to rain penetration. Figure VIII.4 shows a 40-years-old brick wall which is still in a satisfactory condition.

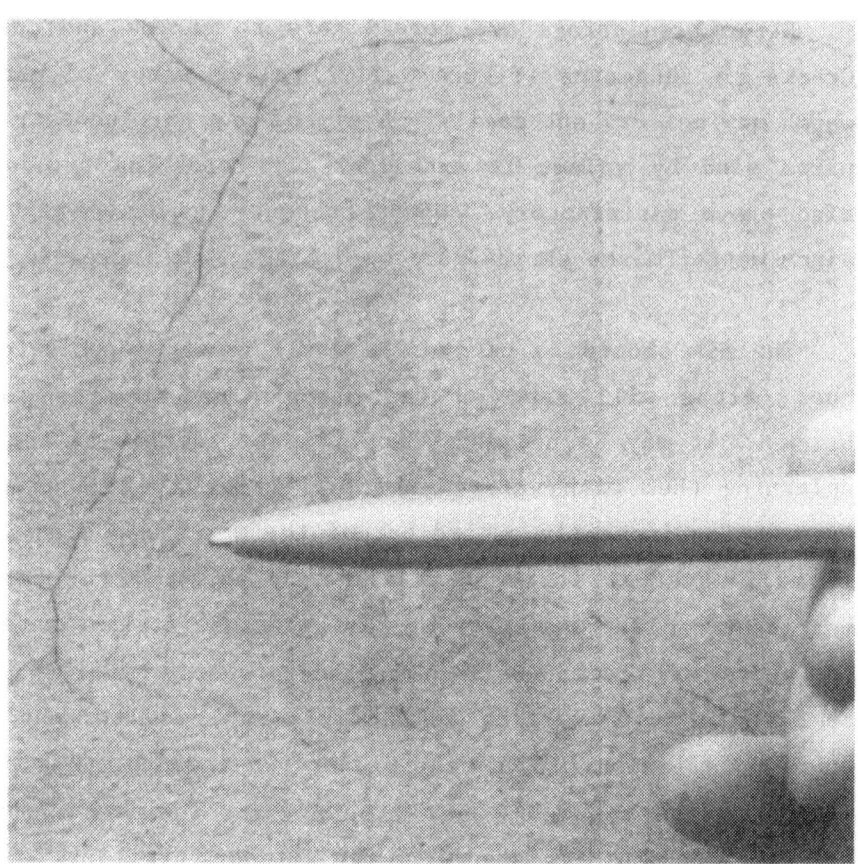

Figure VIII.2

Cracks in rendering (India)

Figure VIII.3

Old brickwork with extra-wide mortar (India)

Figure VIII.4

Bricks and mortar in satisfactory
condition after 40 years of service
(West Sudan)

CHAPTER IX

ORGANISATION OF PRODUCTION

I PRELIMINARY INVESTIGATIONS

A number of factors must first be investigated prior to investing in a brick manufacturing unit. These factors are briefly reviewed below.

A market for the product must first be ascertained. The mere knowledge that more building materials than available are needed at the present time constitutes an insufficient justification for embarking upon a new venture. There must be a reasonable degree of certainty that bricks produced in the early stages of operation of the works do not quickly satisfy a demand which has accumulated over many years. Sustained demand is a prerequisite for profitable operation of the plant.

Future demand for wall-building materials should be investigated by examining housing development proposals, government five-year plans, etc., and by obtaining information from the Housing Ministry and other government or local authorities.

The availability, extent of use, cost and performance of alternative building materials should be examined. The probability of a profitable investment will be greater if bricks constitute a favoured building material for one reason or another (e.g., quick availability, better performance or durability, or lower cost for the finished walls).

The site must be chosen in relation to the available infrastructure and raw materials supply.

The need for permission or licences to operate the brickworks in a chosen location must be investigated. Similarly, information should be obtained on restrictions which may exist on the quantity or maximum depth which can be quarried.

The location of the works should be such that undue nuisance is not caused by noise, smoke, smell, pollution of air or water as regulations may be in force regarding the protection of the environment.

II. INFRASTRUCTURE

There are several major factors which should be considered when planning the establishment of a brickworks. These are briefly described below.

II.1 Site and access

The site should preferably be flat and free from flooding. Whatever the size of the plant, suitable access must be provided to allow the easy transport of materials and output by barrow, bullock cart or lorry. The access road should not be too steep or twisting. Whenever a clay deposit is far from a good track or road, the construction and maintenance cost of a feeder road should be considered in project evaluation.

II.2 Transportation routes

In small works, customers frequently collect the bricks they purchase. Whether this is the case or not, it is advantageous to have good roads leading to the main market areas, especially if the produced bricks must compete against other building materials produced in the same region. Although rail transportation might appear an attractive means of transport, it is less adaptable for delivery to individual construction sites. Furthermore, additional handling of bricks is required, thus increasing not only labour costs but also the chances of bricks breakage.

II.3 Clay

Sufficient clay must be available for the expected life of the works, bearing in mind variation in quality and thickness of the clay deposits. It is preferable to have a larger quantity in reserve in case production is increased at a later date. Whereas a large works might be planned to have a life of 50 years, small-scale works are not usually established for such a long period. While no clear guidance on the life of small-scale brickmaking plants can be provided, it would be sensible to have enough clay in reserve for 20 to 30 years. At the lowest levels of investment, shorter periods may be acceptable.

For bricks 21.5 x 10.3 x 6.5 cm(37), approximately 2 m^3 of clay is used per 1,000 bricks (allowing a 5 per cent drying loss, a 5 per cent firing loss and an overall shrinkage of 10 per cent). Thus, if the clay is dug to a depth of 1 m, 1,000 bricks will require clay from an area of 2 m^2. Consequently, a plant producing 1,000 bricks per day, for 200 days per year and for 25 years will require approximately one hectare of clay deposits.

II.4 Sand

Sand, which may be necessary to reduce shrinkage of a fat clay, can constitute a major item of cost if not available at the brickworks site. If a 20 per cent addition of sand is necessary, a plant producing 1,000 bricks per day will require 3 m^3 of sand (a lorry load) per week.

Alternatively, if it is anticipated that 20 per cent of the bricks will be rejected, the latter could just supply sufficient grog.

II.5 Water

A considerable quantity of water may be required for the preparation of dry-dug clays, the actual amount depending upon the nature of the clay and the forming process. A works producing 1,000 bricks per day and using clay with a 25 per cent moisture content, will require 500 l of water (two oil drums) per day. If this water is to be brought from a distant water source, a small lorry would be required. Extra supplies of water for the wetting of moulds, the cleaning of equipment and washing purposes should be taken into consideration at the project planning stage.

II.6 Fuel

Fuel requirements will depend chiefly on the type, size and operating conditions of the kiln. Estimates of such requirements were already provided in table VII.1. For example, the intermittent scove kiln may require 1 tonne of firewood per 1,000 bricks. Thus, a brickmaking plant operating 200 days per year, and having an expected life of 25 years, would consume the timber produced on nearly 1 km^2 of an African forest(22). Alternatively, the coppicing of a plantation of a slightly larger area (e.g. 1.5 km^2) may provide an equivalent amount of fuel. It may also be noted that some continuous kilns only use one-fifth of the above amount of fuel. Either coppicing or use of fast-growing species, such as eucalyptus, require planning, land and capital. Otherwise, one may not count on a continuous and reliable supply of timber.

Another example relates to coal consumption by a scove kiln. The latter consumes approximately 4 tonnes of coal (a lorry load) per week, for an output of 1,000 bricks per day. The same amount of coal should be sufficient for one month's operation of a Bull's Trench kiln of the same capacity.

Electricity may be essential for some machines and is convenient for many general purposes.

II.7 Labour

Sufficient skilled and unskilled labour should be available within a short distance of the brickmaking plant, especially since the brick kiln requires frequent attention through day and night. In some areas, brickmaking must be temporarily discontinued during the agricultural peak seasons as the workers must return to their farms.

Labour requirements depend upon the type of material used, the degree of mechanisation, the productivity of labour and the quantity and quality of produced bricks. In small-scale, labour-intensive plants, the hand-winning of clay (including removal of overburden and transportation of clay to a nearby brickworks) requires 7 man-hours per tonne (75) or approximately 20 man-hours per 1,000 bricks. Preparation, moulding and firing require another 20 man-hours (5), giving a total of 40 man-hours per 1,000 bricks. Other plants may require 14 to 62 man-hours per 1,000 bricks. A summary of productivity data collected in a number of countries is provided in table IX.1.

Table IX.1
Labour requirements per 1,000 bricks

Production method and quantity (bricks per day)	Location	Man-hours per 1,000 bricks	Reference
Hand-winning and moulding, burning in clamp (2,000)	United Kingdom	40	75,5
Hand-winning, sand-moulding Scotch type kiln (2,000)	Ghana	14	30
Hand-winning, slop-moulding wood fired scove	Lesotho	31	10
Hand-winning, slop-moulding coal-fired clamp	Turkey	32	10
Hand-winning, slop-moulding wood scove	Tanzania	54	10
Hand-winning, sand-moulding wood scove	Sudan	62	10
Hand-winning, soft mud (Berry machine), clamp-firing (8,000)	United Kingdom	19	5
Hand-winning, extrusion, hot floor-drying, Hoffman kiln (20,000)	United Kingdom	15	5
Moderately mechanised brick plant	United Kingdom	5	10

III. **LAYOUT**

III.1 Guidelines

The layout of the works should follow a logical sequence of production stages between the location of the clay deposite and the roadway. Transportation distances within the works, and the amount of handling, should be kept to a minimum. Maximum use should be made of covered areas, where these can be afforded, especially in climates where frequent rain may interrupt or spoil the product.

The alternative sequences of production stages are represented in the form of a flow chart in figure IX.1. Modifications to the chart can be made if deemed necessary. For example, pugged clay may be suitable for shaping without tempering, and further processing stages may be added, such as the option of using a washmill as part of the clay preparation process.

III.2 Example of a general plant layout

The sequence of operations in a small, labour-intensive brickworks producing approximately 2,000 bricks a day with equipment developed at the Intermediate Technology Workshop (United Kingdom) is shown in figure IX.2. Dry, hand-won clay is brought by barrow to pendulum crushers. The fines collected from beneath the sieve are transferred into bins, a layer at a time. Each layer is then watered. The clay is retained by boards set in vertical rebates at the bin fronts. After tempering in the bins overnight or longer, the clay is moved in portable containers, picked up on a forked sack barrow, and set down by each moulder. The table moulds produce stiff green bricks which are set down on edge on small portable shelving units with the help of small pallets. Once they are full, they are moved with the sack barrow and left to dry for a few days. The shelving units are free for re-use, once the bricks reach the leather-hard stage. The bricks are then stacked on the ground for further drying. Fully dried stacks of bricks may then be moved on the sack barrow to the kiln area where they are fired, cooled and prepared for shipping.

The drawing in figure IX.2 is adapted from one produced by ITW. It shows all operations, except winning which is conducted under three separate covers. In practice, the kiln should be positioned in such a manner that smoke and smell are not carried by the wind over the workers. For example, the kiln in figure IX.2 could have been placed further away from the other work stations.

III.3 Required areas for various scales of production

The total area of a given plant will depend on a large number of factors. No estimates of areas will thus be given in this section. Rather, a few illustrative examples are provided below for interested readers. A labour-intensive works in Ghana (32), set up to produce 1,500 bricks per day, covers a total roofed area of 40 m x 18 m. All processes, from clay preparation to hand moulding, drying in racks and firing in a semi-permanent kiln structure, are thus protected against rain. The chimney of the kiln

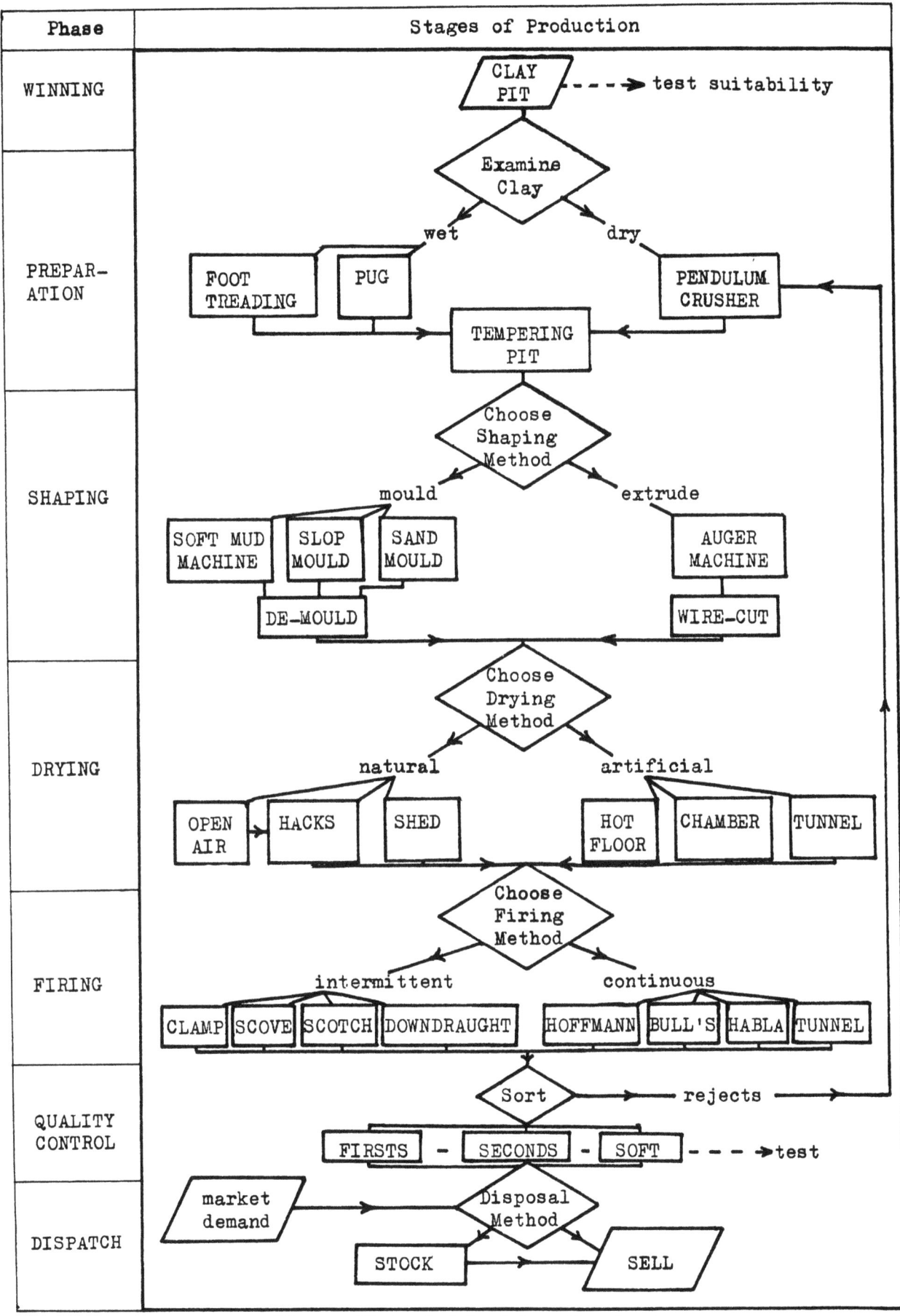

Figure IX.1: Brickmaking flowchart

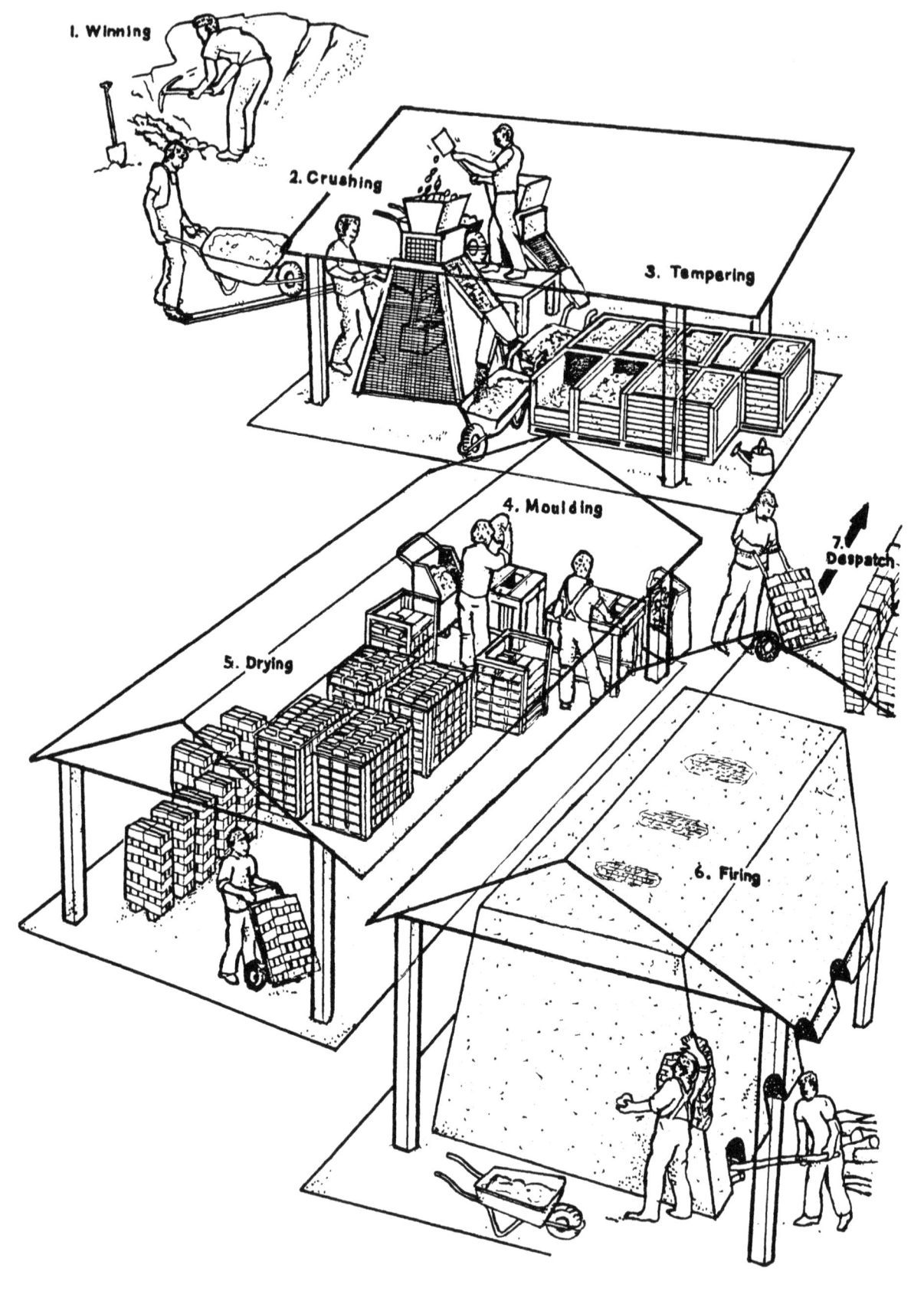

Figure IX.2: Intermediate technology brickmaking plant

penetrates the shed roof. The latter is of great benefit but accounts for approximately half of the capital cost.

Open air drying or drying in covered hacks, with good access between stacks of bricks, requires an area of approximately 1,000 m^2 for an output of 1,000 bricks per day.

An Indian brickmaking plant (35) producing 20,000 bricks per day, and using a washmill, settling ponds, hand moulding, open air drying and a Bull's Trench kiln (60 m x 23 m), covers an area of at least 21,000 m^2. This is at least three times the area used in the Ghana plant per unit of output.

The Bull's Trench kiln brickworks, without a washmill but with an open air drying area, would need twice the area of the small-scale Ghanaian plant per unit of output.

The above examples illustrate extremes in area requirements. Intermediate areas may be required for various types of plants.

Depending on circumstances, sufficient space should also be allocated for offices, sanitary installations and appropriate eating accommodations.

IV. SKILL REQUIREMENTS

The owner or manager of a brickworks should have the necessary skills for managing people as well as equipment and materials. The types of skills necessary for brickmaking are briefly described below.

Digging clay manually or by machine may be thought of as requiring no skill. This is, however, not the case as the careful choice of material at the pit face, and mixing by taking vertical cuts, are essential to good brickmaking. An appreciation of the dip and strike of the clay, and the reasons for rejecting overburden, roots and stones, is desirable. These skills may be taught or obtained from on-the-job training. Care and maintenance of tools should also be part of the skills of pit workers.

Skills are also required for the preparation of the clay for brickmaking, and in appreciating the requirements of a good clay body. Equipment should not be abused and should receive proper maintenance.

Moulding skills are essential to the forming of good bricks. The hand moulder and machine operator must know, for example, that the addition of extra water to the clay, so as to make moulding or extrusion easier, will increase shrinkage and thus the risk of cracking. The hand moulder must have a "feel" for the clay, be able to throw it accurately into the centre of the mould, and produce well-formed bricks at a fast rate over long periods. An understanding of required brick quality will help an operator to maintain his mould or die sizes within the necessary tolerances, allowing for shrinkage on drying and firing. As mechanical devices are likely to wear quickly, checks must be made occasionally, and appropriate maintenance, repair and replacements undertaken whenever necessary.

Those responsible for drying bricks must understand the necessity for careful handling of the latter, especially before they are leather-hard. The operators must know when to turn bricks early in the drying stage (to allow the underneath to dry) and when they are dry enough for firing.

Firing bricks calls for great skill in order to get a good and uniform product. The overall dimension and setting pattern of the brick in building up a kiln, and the spacing between individual bricks in different parts of the kiln, are very important factors. The rate of heating and cooling must be carefully controlled and calls for special skills in the interpretation of various phenomena. Skills are also necessary for the control of the temperature through adjustment of fuel-feeding and draught.

Skilled labour is needed for the sorting of bricks into various grades.

CHAPTER X

METHODOLOGICAL FRAMEWORK FOR THE ESTIMATION
OF UNIT PRODUCTION COSTS

This chapter is intended to assist practising brickmakers wishing to improve their plant and entrepreneurs envisaging to engage in brickmaking by providing them with a methodological framework for the evaluation of alternative brickmaking techniques. Staff of financial institutions, businessmen and government officers may have their own evaluation methodologies, but could still find the following methodological framework useful, especially if they are unfamiliar with brickmaking.

I. THE METHODOLOGICAL FRAMEWORK

The methodological framework consists of two main parts:
(i) the determination of the quantities of various inputs used in the brickmaking process (Step 1 to 5);
(ii) the estimation of the cost for each input, and that of the unit production cost (Step 6 to 12).

These steps are briefly described below. Producers who wish to identify the most appropriate technique should repeat these steps for each technology which may yield the required output.

Step 1: Determination of the number of bricks to be produced each year. This number is a function of market demand, availability of investment funds, the adopted production technique, etc. Chapter IX provides some guidelines for determining the scale of production.

Step 2: Estimation of the quantities of the various materials inputs for the adopted scale of production. The main materials are:

- brickmaking clay
- sand, or an equivalent material (e.g. grog)
- water
- fuel

Guidelines for determining the quantities of each material are given in Chapter IX.

Step 3: Compilation of a list of required equipment, including spare parts and servicing equipment. The list should also include transport equipment whenever necessary as well as testing equipment. Both locally made and imported equipment should be included. Chapters II to VII provide guidelines for the compilation of the above list.

Step 4: Labour requirements. The productivity of the labour force may be significantly different from one country to another. Estimates of labour productivity are given in Chapter IX. These estimates may need to be adjusted for part-time workers. The number of shifts per day, working days per week, and working weeks per year should also be specified, taking into consideration the possible closing down of the plant during the peak agricultural seasons. The number and skills of workers may be established on the basis of the above information.

Step 5: The local infrastructure required must be determined. It may include:
- land for exploitation of the quarry;
- land for access and buildings;
- land for drying grounds and kilns if not within the buildings;
- land for storage of raw materials and of product;
- buildings required, such as covered areas for making, drying or firing; offices; and other amenities.

Chapter IX provide guidelines for the estimation of the area of each of the above items.

Step 6: Working capital. Apart from purchase of land and equipment, it is necessary to have sufficient initial financial resources to enable the purchase of raw materials, including fuel. Working capital will also be required for the payment of wages during an initial period, and for the building up of a stock of bricks for sale.

It is recommended that sufficient working capital be available for two months' materials and fuel, and one month's salaries. If difficulties are anticipated in obtaining supplies of any particular commodity, it may be necessary to hold a stock sufficient for more than two months' production.

Step 7: Equipment and buildings annual depreciation costs. Whatever type of equipment is used, it will have a limited life. An estimate must thus be made of the annual depreciation costs for separate equipment items. The depreciation cost of buildings must also be estimated. These costs will depend upon the initial purchase price, the life of equipment and buildings and the prevailing interest rate. Table X.1 may be used to estimate these depreciation costs. It gives the discount factors (F) for interest rates up to 40 per cent and expected life periods up to 25 years. Thus, if Z is the purchase price of the equipment or the cost of the building, the annual depreciation cost is equal to Z/F.

Hence, the longer the useful life, the lower the annual depreciation cost, and the higher the prevailing interest rate, the higher this cost.

The C.I.F prices of imported pieces of equipment may be obtained from local importers or equipment suppliers (see Appendix IV). The prices of local equipment and buildings may be obtained from local contractors and equipment manufacturers.

Step 8 Land has an infinite life, and the pit may be restored to its original use in some instances. Thus, the annual cost of land may be assumed to be equal to the annual rent of equivalent land. Alternatively, if land is owned by the brickmaker, a hypothetical annual rental rate should be used when estimating the annual land cost.

Step 9: The annual cost of consumable items identified in Step 2 must be calculated. Clay, sand and water are often extremely cheap commodities. The only significant part of their cost, in comparison with that of fuel, is that incurred in extracting and transporting them. Thus, the cost of sand and clay is often included in the cost of labour, land rental rate and equipment depreciation costs. Fuel costs are of major importance and must be determined locally by examining current market prices.

Step 10: The cost of labour varies greatly from one country to another. They must thus be calculated on the basis of local wage rates. The labour requirements are those identified in Step 4.

Step 11: Working capital raised for the project will require an allowance in the annual costs for interest payments to be made on that capital.

Step 12: The total annual cost consists of the sum of the separate annual costs itemised in Steps 7 to 11.

Total annual cost =
Depreciation costs of equipment and buildings +
Annual land rental cost +
Annual labour costs +
Annual fuel costs +
Annual costs of sand and clay (if the latter are purchased rather than won on land belonging (or rented) by the producer) +
Annual cost of water (if the latter is purchased rather than pumped from a river or well within the production unit) +
Annual cost of electricity (if used) +
Annual interest payments on working capital.

The unit production cost of bricks is then equal to:

Total annual cost ÷ Annual output.

The use of the above methodological framework is illustrated in the following example.

II. APPLICATION OF THE METHODOLOGICAL FRAMEWORK

The framework is used to estimate the unit cost of bricks produced by a small-scale brickmaking unit typical of those covered by this memorandum. Production takes place six days per week all year long. The production technique is similar to that described in Section IV.1 of Chapter I.

Step 1 : Annual production of bricks: 624,000 (approximately 2,000 bricks/day)

Step 2 : Annual requirements of clay, sand, water and wood fuel:

Clay : 624 x 2 m^3	= 1,248 m^3
Sand: 20 per cent of clay input	= 250 m^3
Water : 500 l per 1,000 bricks	= 312,000 l
Wood fuel: 1 tonne per 1,000 bricks:	= 624 tonnes

Step 3 : List of equipment:

   Winning : 3 wheelbarrows, picks, shovels

   Preparation: - flat concrete area
   - 2 table moulds (actual mould boxes imported; tables manufactured locally)

   Drying: Mobile rack to hold 4,000 bricks

   Spares: 10 per cent of cost of equipment.

Step 4 : Labour requirements:

Supervision :	1 manager
	1 foreman
Winning:	2 unskilled workers
Preparing:	3 unskilled workers
Shaping:	2 skilled workers
Kiln building and firing:	3 skilled workers (three shifts occasionally worked)
	1 unskilled worker
Carrying, etc.:	1 unskilled worker

Step 5 : Land and building requirements

Land for quarry, dug 2 m deep for 25 years	1.6 ha
Land for access and buildings (including drying)	0.3 ha
Kiln area	0.2 ha
Land for storage of product	0.1 ha
Total land requirement	2.2 ha

Buildings:

covered area	20 x 20 m	400 m^2
office	3 x 3 m	9 m^2
staff amenities	4 x 3 m	12 m^2

Step 6 : Working capital

Working capital to cover cost of materials for 2 months (See step 2) and salaries for 1 month

Clay :	208 m^3
Sand :	42 m^3
Water :	52,000 l
Wood fuel :	104 t
One manager:	1 month
One foreman :	1 month
Five skilled workers :	5 months
Seven unskilled workers :	7 months

Step 7 : Depreciation costs:

For the small plant under consideration, the useful life of the buildings is 25 years. The various tools and pieces of equipment used for production will have much shorter lives. Wheelbarrows, buckets, hand tools, moulds and drying racks, etc. may need replacement after two years.

Annual depreciation costs of the above items are calculated as follows:

Initial costs	Tanzania shillings
- Covered area of 400 m^2 at 50/= per m^2	20,000
- Office and amenities : 21 m^2 at 300/= per m^2	6,300
- Barrows, sundry tools, moulds and racks	4,000
- Spares (10 per cent of tools and equipment cost)	400
Total	30,700

Annual depreciation costs

The discount factor F is equal to (see table X.1):

- 7.843 for buildings with a 25 years, life and a 12 per cent interest rate
- 1.690 for equipment with a 2 years, life and a 12 per cent interest rate

The annual depreciation cost of buildings is then equal to:

$$\frac{26,300/=}{7.843} = 3.350/=$$

and that of equipment is equal to

$$\frac{4,400/=}{1.690} = 2,600/=$$

Total annual depreciation costs are therefore equal to:

$$3,350/= + 2,600/= = \underline{5,950/=}$$

Step 8 : Annual rental rate of land:

Land utilised for brickmaking is likely to be situated in areas commanding low land value or rental. If brickworks are known to be operating in similar land, rentals paid for the latter may serve as a guide to prospective brickmakers. Otherwise, agricultural land price or rentals may be used for comparative purposes.

The annual rental rate in this example is assumed to be 500 T. Shillings.

Step 9 : Annual cost of materials:

			Tanzania Shillings
- clay	1,248 m^3 at 0.72/m^3	=	90
- sand	250 m^3 at 0.12/m^3	=	30
- water	320 m^3 at 0.385/m^3	=	120
- wood fuel	624 tonnes at 60/tonne	=	37,440
Total annual cost of materials		=	37,680

Step 10: Annual labour costs:

		Tanzania Shillings
1 manager at 450/= per month	=	5,400
1 foreman at 400/= per month	=	4,800
5 skilled workers at 250/= per month each	=	15,000
7 unskilled workers at 200/= per month each	=	16,800
Total annual labour cost	=	42,000

Step 11 : Interest payments on working capital:

The cost of materials and labour listed in step 6 is equal to 9,780/=, given the unit prices indicated in steps 9 and 10.

The annual interest payments are therefore equal to:

9,780/= X .12 = 1,170 Tanzania Shillings

Step 12 : Total unit production cost:

The total annual production cost is equal to:

5,950/= + 500/= + 37,680/= + 42,000/= + 1,170/= = 87,300 Tanzania Shillings

The unit production cost of bricks is then equal to:

87,300/= ÷ 624,000/= = 0.14/= (14 cents)

- 176 -

Year (X)	Interest Rate (I)															
	5%	6%	8%	10%	12%	14%	15%	16%	18%	20%	22%	24%	25%	26%	28%	30%	35%	40%
1	0.952	0.943	0.926	0.909	0.893	0.877	0.870	0.862	0.847	0.833	0.820	0.806	0.800	0.794	0.781	0.769	0.741	0.714
2	1.859	1.833	1.783	1.736	1.690	1.647	1.626	1.605	1.566	1.528	1.492	1.457	1.440	1.424	1.392	1.361	1.289	1.224
3	2.723	2.673	2.577	2.487	2.402	2.322	2.283	2.246	2.174	2.106	2.042	1.981	1.952	1.923	1.868	1.816	1.696	1.589
4	3.546	3.465	3.312	3.170	3.037	2.914	2.855	2.798	2.690	2.589	2.494	2.404	2.362	2.320	2.241	2.166	1.997	1.849
5	4.330	4.212	3.993	3.791	3.605	3.433	3.352	3.274	3.127	2.991	2.864	2.745	2.689	2.635	2.532	2.436	2.220	2.035
6	5.076	4.917	4.623	4.355	4.111	3.889	3.784	3.685	3.498	3.326	3.167	3.020	2.951	2.885	2.759	2.643	2.385	2.168
7	5.786	5.582	5.206	4.868	4.564	4.288	4.160	4.039	3.812	3.605	3.416	3.242	3.161	3.083	2.937	2.802	2.508	2.263
8	6.463	6.210	5.747	5.335	4.968	4.639	4.487	4.344	4.078	3.837	3.619	3.421	3.329	3.241	3.076	2.925	2.598	2.331
9	7.108	6.802	6.247	5.759	5.328	4.946	4.772	4.607	4.303	4.031	3.786	3.566	3.463	3.366	3.184	3.019	2.665	2.379
10	7.722	7.360	6.710	6.145	5.650	5.216	5.019	4.833	4.494	4.192	3.923	3.682	3.571	3.465	3.269	3.092	2.715	2.414
11	8.306	7.887	7.139	6.495	5.938	5.453	5.234	5.029	4.656	4.327	4.035	3.776	3.656	3.544	3.335	3.147	2.752	2.438
12	8.863	8.384	7.536	6.814	6.194	5.660	5.421	5.197	4.793	4.439	4.127	3.851	3.725	3.606	3.387	3.190	2.779	2.456
13	9.394	8.853	7.904	7.103	6.424	5.842	5.583	5.342	4.910	4.533	4.203	3.912	3.780	3.656	3.427	3.223	2.799	2.468
14	9.899	9.295	8.244	7.367	6.628	6.002	5.724	5.468	5.008	4.611	4.265	3.962	3.824	3.695	3.459	3.249	2.814	2.477
15	10.380	9.712	8.559	7.606	6.811	6.142	5.847	5.575	5.092	4.675	4.315	4.001	3.859	3.726	3.483	3.268	2.825	2.484
16	10.838	10.106	8.851	7.824	6.974	6.265	5.954	5.669	5.162	4.730	4.357	4.033	3.887	3.751	3.503	3.283	2.834	2.489
17	11.274	10.477	9.122	8.022	7.120	6.373	6.047	5.749	5.222	4.775	4.391	4.059	3.910	3.771	3.518	3.295	2.840	2.492
18	11.690	10.828	9.372	8.201	7.250	6.467	6.128	5.818	5.273	4.812	4.419	4.080	3.928	3.786	3.529	3.304	2.844	2.494
19	12.085	11.158	9.604	8.365	7.366	6.550	6.198	5.877	5.316	4.844	4.442	4.097	3.942	3.799	3.539	3.311	2.848	2.496
20	12.462	11.470	9.818	8.514	7.469	6.623	6.259	5.929	5.353	4.870	4.460	4.110	3.954	3.808	3.546	3.316	2.850	2.497
21	12.821	11.764	10.017	8.649	7.562	6.687	6.312	5.973	5.384	4.891	4.476	4.121	3.963	3.816	3.551	3.320	2.852	2.498
22	13.163	12.042	10.201	8.772	7.645	6.743	6.359	6.011	5.410	4.909	4.488	4.130	3.970	3.822	3.556	3.323	2.853	2.498
23	13.489	12.303	10.371	8.883	7.718	6.792	6.399	6.044	5.432	4.925	4.499	4.137	3.976	3.827	3.559	3.325	2.854	2.499
24	13.799	12.550	10.529	8.985	7.784	6.835	6.434	6.073	5.451	4.937	4.507	4.143	3.981	3.831	3.562	3.327	2.855	2.499
25	14.094	12.783	10.675	9.077	7.843	6.873	6.464	6.097	5.467	4.948	4.514	4.147	3.985	3.834	3.564	3.329	2.856	2.499

TABLE X.1: PRESENT WORTH OF AN ANNUITY FACTOR (F)

CHAPTER XI

SOCIO-ECONOMIC IMPACT OF ALTERNATIVE
BRICK MANUFACTURING TECHNIQUES

The social and economic effects of various brickmaking options should be of particular interest to public planners, financial institutions, business men, housing authorities and project evaluators from industrial development agencies. The purpose of this chapter is therefore to shed light on these various effects with a view to helping public planners identify technologies consonant with national socio-economic objectives, and formulate appropriate policies for the promotion of these technologies.

I. EMPLOYMENT GENERATION

Employment generation constitutes one major objective of national development plans of developing countries. Thus, technologies which require more labour per unit of output than other technologies should be favoured as long as labour is used in an efficient manner. Consequently, small-scale brickmaking techniques should be favoured - from an employment viewpoint - over those used in large-scale turnkey factories as they do generate substantially more employment than the latter technologies. For example, a study of brickmaking techniques in Bogotà (Colombia) shows that large-scale brick manufacturing plants produce 83 per cent of the national brick production although they use only one-third of the total labour force in the brick industry. In other words, small-scale units, using relatively labour-intensive techniques, require approximately ten times more labour than large-scale plants per unit of production (76). Another study carried out in Colombia (43) shows that the labour-capital ratio in small-scale brickmaking units is over 100 times greater than that for large-scale manufacturing plants.

Table XI.1 provides estimates of labour inputs per 10 million bricks for various scales of production and techniques. It may be seen that the small-scale plants require between 20 and 25 times more labour than the highly automated plants for the same output.

Table XI.1

Scales of production and labour generation

Description of brickmaking methods	Production rate of one plant (bricks per day)	Labour per 10 million bricks per year
Small-scale, traditional manual processes	2 000	160
Small-scale, intermediate technology	2 000	200
Soft-mud machine, otherwise manual	14 000	76
Moderately mechanised	64 000	20
Highly automated	180 000	8

II. TOTAL INVESTMENT COSTS AND FOREIGN EXCHANGE SAVINGS

The import of expensive equipment with subsequent need for spare parts and services can be a burden on a country's foreign exchange reserves. The choice of small-scale techniques and the use of locally manufactured equipment, which can be repaired by local craftsmen, may thus partly alleviate the need for imports and scarce foreign exchange.

A large number of developing countries also suffer from shortages of local capital funds as exemplified by the high interest rates prevailing in these countries. Consequently, technologies which minimise the use of capital investments should be favoured.

Table XI.2 shows that, from a foreign exchange and capital investment viewpoint, small-scale brickmaking technologies are, by far, much more appropriate than large-scale, automated plants. Thus, capital investments in large-scale plants are six to 100 times larger than those for small-scale plants. Similarly, the import component of these investments is five to 15 times larger for the automated plants than for small-scale units.

Table XI.2

Capital costs and foreign exchange inputs

Description of brickmaking processes	Total cost for 10 million bricks per year ('000 US$)	Proportion of costs (per cent)	
		Import	Local
Small-scale, traditional manual process	34	5	95
Small-scale intermediate technology	578	15	85
Mechanical plant with Hoffman kiln	3,880	75	25

Source: 77

III. UNIT PRODUCTION COST

A major component of the cost of a house in developing countries is that of building materials. This is particularly the case for low-cost housing or various self-help schemes. It is therefore important to promote the production of low-cost building materials in order to facilitate home ownership by low-income groups and to reduce public investments by housing authorities. In the case of bricks, production techniques and scales of production which minimise unit production costs should therefore be favoured. Although reliable estimates of these costs are not available, recent studies (78) tend to indicate that small-scale plants, using intermediate technologies, produce bricks at a lower cost than large-scale, capital-intensive plants. Results of these studies are summarised in table XI.3. It may be seen that the unit price of bricks produced in small-scale plants is two to three times lower than that of bricks produced in large-scale plants.

Table XI.3

Unit production costs

Classification of brickmaking process	Unit production cost (US cents per brick)	
	Medium wage regime	Low wage regime
Capital-intensive, all year round	6.5	6.2
"Least-cost", all year round	3.1	2.3
"Least-cost", seasonal working only	2.9	2.0

Source: 78

IV. RURAL INDUSTRIALISATION

It is desirable to set up industries in rural areas, to counterbalance the faster growth rate of urban areas (2). These rural industries, however, must be of a size appropriate to the local markets and to the social needs of the people. It should particularly be noted that the higher the degree of automation (or of mechanisation) the higher the required professional qualifications (7). Consequently, the chosen level of technology should be such as to ensure that any maintenance or repair can be carried out on site. Small-scale plants, which can operate without electricity, are also particularly suitable for rural areas.

It is also essential that the adopted technology be consonant with the qualification level of the local labour force. Some aspects of skill requirements for brickmaking have been discussed in Section IV of Chapter IX. These requirements do not constitute a major constraint to the establishment of rural brickmaking units, especially if equipment is locally produced and repaired. Furthermore, small-scale production techniques should be flexible enough so that production may be increased according to market requirements (30). The brickmaking plants may also have to operate on a seasonal basis (according to weather conditions and the availability of workers during peak agricultural seasons), a condition which overrules the adoption of large-scale, capital-intensive plants in rural areas.

V. MULTIPLIER EFFECTS

Alternative production techniques will have different multiplier effects on the national economy (i.e. backward and forward linkages) depending on the scale of production and the origin of equipment and materials. From a socio-economic viewpoint, technologies associated with the largest multiplier effects should, all other things being equal, be favoured over those with limited effects.

Small-scale, relatively labour-intensive brickmaking technologies are generally associated with larger multiplier effects than large-scale, capital-intensive technologies for the following reasons. Firstly, most, if not all, the equipment (e.g. table moulds, mixing equipment) may be manufactured locally instead of being imported (see section II). Production of such equipment should thus generate additional employment and incomes. Secondly, small-scale plants use local fuel (e.g. wood, agricultural wastes) while large-scale plants often rely on oil. The gathering and processing of fuel wood or other local materials will thus generate additional employment. Finally, the marketing and transport of bricks produced by small-scale units should also generate more employment than those of the output of large-scale plants. Altogether, employment generated through multiplier effects will further expand that generated through the production of bricks.

VI. ENERGY REQUIREMENTS

Fuel requirements of various types of kilns have already been considered in Chapter VII. The apparent higher fuel efficiency of large kilns must be weighed against the fact that large plants operate with a considerable amount of energy-consuming equipment. For example, a large plant producing 10 million bricks per year may require 300 kWh electricity supply for the running of the equipment(7). Furthermore, transportation costs are likely to be higher in large-scale plants than in small-scale plants. Thus, the low fuel efficiency of small-scale kilns is partly offset by the higher energy inputs for equipment used in large-scale plants. However, as energy requirements are partly a function of the efficiency of kiln-operating procedures, it is difficult to compare the relative fuel efficiency of small and large production units.

VII. CONCLUSION

Small-scale production of bricks is generally preferable to large-scale production under the following circumstances: where low wage regimes exist; where foreign exchange is in short supply; where no well-developed industrial base exists; where infrastructure and technical skills are insufficient; and where the market is small or widespread.

Labour-intensive brickmaking techniques are often considered the most appropriate techniques under conditions prevailing in developing countries. However, recent studies (78) show that this may not always be the case if the factors mentioned above are not present. Thus, project evaluators should properly consider all available alternatives prior to promoting one technology or another. The evaluation of alternative technologies should take into consideration the fact that small-scale production generates more employment than large-scale production while minimising capital investments and foreign exchange costs.

APPENDICES

APPENDIX I

GLOSSARY OF TECHNICAL TERMS

Adobe	Mud brick; hand made, dried in the sun, not fired.
Alluvial material	Clay, sand or mud laid down by the flooding of a river.
Arris	The edge where two clay faces meet.
Auger	Tool for boring a hole in the ground or taking a sample of the soil; it has a screw-like action. Also a machine which forces clay through an aperture by means of a screw thread rotating inside a barrel containing the clay.
Bag wall	The brick wall built around the back of each of the fires of a downdraught kiln.
Batter	To slope the face of an embankment or quarry; opposite of overhand.
Bed face	The underneath surface of a brick as laid in a wall, usually the largest face; the face of a brick bedded in the mortar.
Benching	Method of winning clay from the pit simultaneously at several different levels, by working at several benches or steps.
Binder	The material which binds together separate particles; for example, cement and lime, used to make mortars, are binders.

Blade	A thin section of brick across the whole width of a kiln.
Bloating	Creation of gas bubbles within the near-vitrified clay in the kiln, causing blisters and craking on brick surfaces.
Block	A building unit larger than a brick, usually requiring two hands to lift it.
Body	The material after processing.
Bond	The pattern of arrangement of bricks in a wall, usually such that vertical joints between bricks are not immediately above each other in adjoining courses.
Brick	A unit from which walls may be built and which is of such size and weight that it can be laid with one hand, allowing the other hand to be used for operating with a trowel.
Bulking of sand	A given weight of sand will have various volumes, depending upon the water content.
Bull's trench kiln	An archless continuous kiln based on the principle of the Hoffmann kiln.
Burning	Firing. In the case of bricks, burning them changes the nature of the clay from which they have been shaped, increasing their strength and durability.
Calcium silicate	A generally durable compound formed when lime reacts with silica, as for example during the high pressure and high temperature methods of treating lime and silica sand to make calcium silicate bricks.
Centring	Wooden or other formwork built up to support a brick arch while the mortar sets.

Clamp	Large pile of green bricks with fuel between those in the bottom courses, which is set on fire to burn the bricks. Sometimes, this term is also used to describe a similar pile but with the fuel placed in tunnels through the lower courses of the pile.
Continuous kiln	A kiln in which the fire is always burning, bricks being warmed, fired, and cooled simulatenously in different parts of the kiln.
Coppicing	The annual cutting of wood from trees, which therefore do not grow to full size, yet continue to provide useful materials such as fuel.
Course	A horizontal layer of bricks.
Cuckhold	A concave-bladed spade for cutting off lumps of prepared clay.
Cuckle	A curved metal strip with two handles for cutting off a pice of clay from a large lump.
Dip	The slope of a clay deposit compared with the horinzontal.
Docking	Immersing fired bricks in water for a short while so that a thin outer skin is thoroughly wetted. Claimed by several authorities to reduce the incidence of lime blowing.
Downdraught kiln	A kiln in which hot gases pass down between the bricks which are being fired.
Ettringite	A compound formed from sulphates and calcium aluminate; if it is formed after the mortar between the bricks is well set, it may cause softening and expansion of the mortar joints.

Eye	The point near the base of a clamp where the fire is first lit.
Fair-faced brickwork	Brick walling of an acceptable standard of appearance and quality, without rendering or plastering.
Fat soil	Highly plastic soil; usually a clay rich soil with high drying shrinkage.
Firing	Heating in a kiln to partially vitrify (see burning).
Flash	A silver of clay on the arris of a brick, formed in the crack or air inlet in the mould.
Flash wall	A long wall behind the fires on one side of a down-draught kiln, serving to deflect the hot gases upward.
Flux	A mineral in the clay which reduces the temperature required to obtain vitrification.
Frog	Indentation in one or sometimes both of the bed faces of a brick. A frog cannot be put into an extruded, wirecut brick, but is easily formed in moulded or pressed bricks.
Ghol	Clay washing tank (Indian).
Green brick	Brick formed into shape but not yet fired.
Grog	Fired clay, often reject bricks, crushed to a fine size, for addition to the clay body. Such material reduces shrinkage and opens the body.
Habla kiln	An archless zigzag continuous kiln, based on the Hoffmann kiln.
Hard fired bricks	Bricks fired to a relatively high temperature, producing a moderate amount of vitrification and consequent good strength and durability.

Header	The small end of a brick showing in the face of brickwork.
Heat work	The combination of temperature and time and its effect on ceramic reactions.
Hoffmann kiln	A brick-arched continuous kiln, circular or elliptical, in which waste heat is used to preheat both the combustion air and the bricks, so increasing fuel efficiency.
Hydraulic lime	A lime which will set under water. This is due to cetain siliceous impurities which react with the lime itself. Non-hydraulic limes harden only by carbonation caused by carbon dioxide in the air.
Igneous rocks	Rocks of volcanic origin; rocks which were molten at one time.
Intermittent kiln	A kiln in which the fire is allowed to die out and the bricks to cool after they have been fired. The kiln must be emptied, refilled and a new fire started for each load of bricks.
Lean soil	Low plasticity soil, usually due to lack of the finer sizes of clay fractions. In contrast to fat soil.
Leather-hard bricks	Bricks which have partly dried so that they can be picked up without distorting them.
Loam	Sandy clay often suitable for shaping into bricks, and having a low drying shrinkage.
Marl	Natural mixture of clay and chalk.

Open the clay body	To increase the permeability to gases of a ceramic body.
Overburden	The material lying on top of a natural deposit of brick clay.
Oxidising	Conditions of surroundings in which oxygen is freely available.
PCE	Abbreviation for pyrometric cone equivalent.
Perpends	The visible vertical joints between bricks in a wall.
Plastic material	Material able to be deformed by moderate pressure and retaining the deformed shape when the pressure is removed.
Plasticity	Possessing the property of being plastic.
Profile	A section taken through the various strata of a soil.
Puddle	To mix up dry soil or lime with water, usually manually.
Pyrometric cone	A small clay-based cone which will squat after undergoing a certain amount of heat work.
Reducing	Conditions of surroundings in which little or no oxygen is available.
Refractory	Heat-resisting.
Render	To cover an exterior wall surface with a cement-lime based mix.
Ring	The high-pitched metallic sound obtained when two well-fired bricks are struck against each other.

Runner	Person who carries slopmoulded bricks in mould to the drying ground.
Scove	To cover over with mud. The name of the scove kiln originates from the practice of scoving the outside bricks in order to stop the heat from escaping from the pile of bricks being fired.
Sesquioxides	The oxides of aluminium and iron.
Shale	Soft laminated slate-like rock, harder than most clays.
Short material	Lacking plasticity, lean.
Siliceous	Having a high proportion of silica.
Slake	To fall apart when immersed in water.
Slip	A thin slurry of clay in water; very wet and runny mix of clay.
Soak stage	Period during which bricks are kept at a fixed elevated temperature in a kiln. Also immersion of bricks in water.
Soft-fired bricks	Bricks heated in a kiln to a relatively low temperature; bricks so treated do not exhibit optimal physical properties.
Solar gain	Heat obtained from the sun.
Sour	Leave clay in contact with water for a long period.
Spall	Flake away from the surface.
Specific surface area	The total area of either the many fine particles or the many fine pores in a solid within a standard weight of the material.

Squat	The deformation of a clay near its vitrification point, especially the deformation of a pyrometric cone.
Strain	A measure of the change in size compared to the original size.
Strata	The various layers in a sedimentary deposit.
Stress	A measure of the force applied to produce strain in an object
Stretcher	The long face of a brick (not the bed face) showing in a wall.
Strike	The direction in a clay deposit in which the clay is at the same depth. Also a piece of wood for pushing off excess clay in slop moulding.
Striker	Piece of wood for pushing off excess clay in slop moulding.
Sump	A depression into which water may be drained off, or wastes deposited.
Surkhi	Soft-fired clay, ground up, for mixing with lime to make mortar (Indian).
Temper	Leave in wet condition, often overnight or longer to make clay more workable and easier to mould.
Terracing	Benching.
Thermal capacity	A measure of the quantity of heat which an object can hold; high values in building components reduce temperature extremes within the building.
Tunnel kiln	Closed kiln or dryer through which the bricks are carried on wheeled cars.

Updraught kiln	A kiln in which the hot combustion gases pass upward through the bricks which are being fired.
Water-smoking	The first stages of heating in a kiln during which only a gentle heat is applied to remove remaining water from the green bricks.
Wicket	The doorway for access into a kiln. It is bricked up temporarily whilst the bricks are being fired.
Winning	Obtaining raw material from a deposit.
Worked out	Description applied to a deposit which has been completely dug.

APPENDIX II

BIBLIOGRAPHICAL REFERENCES

(1) Habitat: Global review of housing, conference on Human Settlements, Doc. No. A/CONF/70/A1 (Vancouver, 1976).

(2) United Nations Commission on Human Settlements: Report on Fifth Session, Habitat News (Nairobi), Vol. 4, No. 2, Aug.-Sep. 1982, p. 10.

(3) Sikander, A.S.; Quadeer, M.A. : "Squatter settlements - A functional view", in Vol. 2 of the Proceedings of a conference of the International Association for Housing Science held in Dharan in 1978, (New York, Wiley and Sons, 1982), pp. 437-446.

(4) UNIDO: Development of clay building materials industries in developing countries, report of seminar held in Copenhagen, Doc. No. ID/28 (Vienna, 1968).

(5) UNIDO: Establishment of the brick and tile industry, Doc. No. ID/15 (Vienna, 1969).

(6) UNIDO: Brickmaking plant. Industry profile. Development and transfer of technology series No. 10 (New York, 1978).

(7) Bender, W.: The planning of brickworks (Plymouth, Macdonald and Evans, 1978).

(8) Searle, A.B.: Modern brickmaking (London, Ernest Benn, 1956).

(9) UNIDO: Clay building materials industries in Africa, report of a workshop held in Tunis in 1970 (Vienna, 1971).

(10) Parry, J.P.M.: Brickmaking in developing countries (Garston, Watford, British Research Institute, 1979).

(11) Lunt, M.G.: Stabilised soil blocks for buildings. Overseas Building Note No. 184 (Garston, British Research Institute, 1980).

(12) Smith, R.G. : "Small-scale production of gypsum plaster for building in the Cape Verde Islands" in Appropriate Technology (London, Intermediate Technology Development Group) 1982, Vol. 8, No. 4, pp.4-6.

(13) _____ : Long-term unrestrained expansion of test bricks in Transactions of the British Ceramic Society (London), 1973, Vol. 72, No. 1, pp. 1-5.,.

(14) British Research Institute: "Materials for concrete" in Digest No. 237 (Garston, Watford), 1980, p. 237.

(15) _____ : "Repairing brickworks" in Digest No. 200 (Garston), 1981, p. 2.

(16) Schumacher, E.F.: Small is beautiful: A study of economics as if people mattered (London, Blond and Briggs, 1973).

(17) Grimshaw, R.W.: The chemistry and physics of clay, (London, Ernest Benn, 1971).

(18) British Standards Institution: Methods of test for soils for engineering purposes Doc. No. BS1377 (London, 1975).

(19) Doat, P.; Hay, A.; Houben, H.; Matuk, S.; and Vitoux, F.: Construire en terre (Paris, Collection An-Architecture, 1979).

(20) Stulz, R.: Appropriate building materials. Publication No. 12 (St. Gall, Swiss Centre for Appropriate Technology 1981).

(21) Department of Housing and Urban Development, Office of International Affairs: Handbook for building houses of earth, (Washington, DC, n.d.).

(22) Knizek, I.: Brickmaking plant. Industry profile. Development and transfer of technology series No. 10 (New York, UNIDO, 1978).

(23) Butterworth, B.: Methods of assessing the suitability of clays for brickmaking (London, Claycraft, 1947).

(24) Clews, F.H.: Heavy clay technology, (Stoke-on-Trent, British Ceramic Research Association, 1969).

(25) West, H.W.H.: Production technology - Winning, preparation and shaping of clay, Doc. No. ID/WG/81/4 (New York, UNIDO, 1970).

(26) Smith, R.G.: Brickmaking by Malagasy artisans and the establishment of a pilot centre (Geneva, ILO, 1980).

(27) Sedalia, B.M.: Structural clay industry (Bandung, United Nations Regional Housing Centre, 1976).

(28) Svare, T.I.: Better burnt bricks, technical pamphlet No. 1 (Dar-es-Salaam, National Housing and Building Research Unit, 1971).

(29) Harley, G.T.: "A study of shovelling" in Transactions of the British Ceramics Society (Stoke-on-Trent), 1932, Vol. 31, No. 1, pp. 1-35.

(30) Chatterjee, A.K.: "Clay preparation and product manufacture", in Small-Scale Building and Road Research News (Khumasi), 1977, Vol. 2, No. 2.

(31) Srinivasan S. and Jain L.C.: "Lime bursting in bricks" in Digest No. 113 (Roorke, India, Central Building Research Institute, 1975).

(32) Amonoo-Neizer, K.: Asokwa brick project, Special Report No. SR 1/73 (Kumasi, Ghana, Building and Road Research Institute, 1973).

(33) Woodforde, J.: Bricks to build a house (London, Routledge and Kegan Paul, 1976).

(34) Hammond, M.: Bricks and brickmaking (Princes Risborough, United Kingdom, Shirs Publications, 1981).

(35) Prakash, S.; Majundar, N.C.: "Manufacture of bricks of improved quality in Bhopal", in Journal of Engineers and Planners (New Delhi), 1974, Vol. 2, Nos. 8 & 9, pp. 27-30.

(36) Indian Standards Institution: Specification for burnt clay facing bricks, Doc. No. IS 2691-1972 (New Delhi, 1972).

(37) British Standards Institution: Specification for clay bricks and blocks, Doc. No. IS No. 3921-1974 (London, 1974).

(38) Majumdar, N.C.; Wadhwa, S.S.; Hiralal, E.S.: "Manufacture of building bricks by a semi-mechanised process", in Transactions of the Indian Ceramics Society (New Delhi), 1969, Vol. XXVIII, No. 4, pp. 121-128.

(39) Majumdar, N.C.; Hiralal, E.S.; Handa, S.K.: An appropriate technology for mechanised production of building bricks. Proceedings of a national seminar on building materials - their science and technology (New Delhi, 1982).

(40) Thomas, D.W.: Small-scale manufacture of burned building brick (Arlington, Virginia, Volunteers in Technical Assistance, 1977).

(41) Smith, R.G.: "Improved moulding devices for hand-made bricks", in Appropriate Technology (London, Intermediate Technology Publications), 1981 Vol. 7, No. 4.

(42) Weller, H.O.; Campbell, A.J.: Brickmaking in East Africa (Nairobi, East African Industrial Research Board, 1945).

(43) Baily, M.A.: "Brick manufacturing in Colombia: A case study of alternative technologies", in World Development (London, Pergamon Press), 1981, pp. 201-213.

(44) Ford, R.W.: Drying, Institute of Ceramics Textbook series No. 3 (London, Maclaren, 1974).

(45) Macey, H.H.: Drying in the heavy clay industry, National Brick Advisory Council Paper No. 3 (London, Her Majesty Stationary Office, 1950).

(46) Small Industries Development Organisation: Burnt clay brickmaking, Rural Industries Guide No. 4 (Dar-es-Salaam, n.d.).

(47) Schmidt, H.: "Measures to counteract defects in bricks during firing in Ziegelindustrie International, Issue No. 3, Mar. 1980, pp. 153-162.

(48) Noble, W.: The firing of common bricks, National Brick Advisory Council Paper No. 4 (London, Her Majesty Stationary Office, 1950).

(49) Jain, L.C.: "Effect of sodium chloride on the prevention of lime blowing" in Indian Ceramics (New Delhi), Mar. 1980, Vol. 17, No. 7, pp. 262-266.

(50) Laird, R.T.; Worcester, M.: "The inhibiting of lime blowing in bricks" in Transactions of the British Ceramic Society, 1956, Vol. 55, No. 8.

(51) Spence, R.J.S.: An investigation of the properties of rural and urban bricks, Paper No. TR9 (Lusaka, National Council for Scientific Research, 1971).

(52) Gundry, D.G.: "Brickmaking on the farm" in Rhodesia Agricultural Journal (Harare), 1951, Vol. XLVIII, No. 4,, pp. 330-343.

(53) Hill, N.R.: "A clamp can be appropriate for the burning of bricks", in Appropriate Technology, 1980, Vol. 7, No. 1,.

(54) Majumdar, N.C.: "Firing of Bull's Trench kilns", in Indian Builder, Sep. 1957.

(55) Indian Standard Institution: Guide for the design and manufacture of brick kilns, Doc. No. IS 4805-1968 (New Delhi, 1968).

(56) Spence, R.J.S.: "Brick manufacture using the bull's trench kiln", in *Appropriate Technology*, 1975, Vol. 2, No. 1.

(57) "New archless continuous kiln", in *British Clayworker* (London), May 1929.

(58) Majumdar, N.C.; Hiralal, E.S.: "High draught kiln : its operation, control and economics", in *Brick and Tiles News* (Roorke, India, Central Building Research Institute), 1980, pp. 47-51.

(59) Salmang, H.: *Ceramics: physical and chemical fundamentals* (London, Butterworths, 1961).

(60) FAO: *Yearbook 1981* (Rome, 1982).

(61) Madibbo, A.M.; Richter, M.: "Fired clay bricks in the Sudan", in *Building Research Digest* (Khartoum, National Building Research Station), 1970, No. 6, Phase 1.

(62) Rad, P.F.: "A simple technique for determining strength of brick", in *Proceedings of the North American Masonry Conference* (1978) Part. 40, pp. 1-10.

(63) Jain, L.C.: "Accelerated test for lime blowing", in *British Clayworker*, 1971, Vol. 80, No. 947, pp. 40-41.

(64) Butterworth, B.: "The frost resistance of bricks and tiles - A review", in *Journal of the British Ceramic Society* (Stoke-on-Trent), 1964, Vol. 1, No. 2, pp. 203-223.

(65) Spence, R.J.S.: *Small-scale production of cementitious materials* (London, Intermediate Technology Publications, 1980).

(66) Smith, R.G.: *Rice husk ash cement* (Rugby, United Kingdom, Intermediate Technology Industrial Service, 1983).

(67) Central Building Research Institute: *Cementitious binder from waste lime sludge and rice husk*, Technical Note No. 72, (Roorke, India, 2nd edition, 1980).

(68) Spence, R.J.S.: *Alternative cements in India* (London, Intermediate Technology Development Group, 1976).

(69) Indian Standards Institution: *Specification for lime pozzolana mixture*, No. IS 4098-1967 (New Delhi, 1967).

(70) Beningfield, N.: *Aspects of cement-based mortars for brickwork and blockwork concrete* (London, 1980).

(71) Smith, R.G. : *Gypsum*, Proceedings of a meeting on small-scale manufacture of cementitious materials (London, Intermediate Technology Publications, 1974).

(72) British Standards Institution: *Specification for masonry cement*, Doc. No. BS 5224-1976 (London, 1976).

(73) _____.: *Code of practice for structural use of masonry*, Part 1 : *Unreinforced Masonry*, Doc. No. BS 5628 Pt1-1978 (London, 1978).

(74) Buttterworth, B. : *The properties of clay building materials*, Paper presented to a Ceramics Symposium (Stoke-on-Trent, British Ceramic Society, 1953).

(75) Macey, H.H.; Green, A.T.: The labour involved in making and firing common bricks, National Brick Advisory Council Paper No. 2 (London, HMSO, 1947).

(76) Centro Nacional de la Construccion: Diagnosis of the economic and technological State of the Colombian brickmaking industry, Doc. No. CEN 10-76 (Bogota, 1976).

(77) Parry, J.P.M.: Technical options in brick and tile production, Paper presented to an Intermediate Technology Workshop (Birmingham, 1983).

(78) Keddie, J.; Cleghorn, W.: "Least cost brickmaking", in Appropriate Technology, 1978, Vol. 5, No. 3, pp. 24-27.

(79) British Research Establishment: Building research centres and similar organisations throughout the world, Overseas Building Note No. 163 (Garston, Watford, 1978).

(80) UNIDO: Information sources on the ceramics industry, Guide to Information Sources No. 17 (New York, 1975).

APPENDIX III

INSTITUTES FROM WHERE INFORMATION CAN BE OBTAINED

ARGENTINA
Association Tecnica Argentina de Ceramica,
Talcahuano 847,
P.B. Buenos Aires.

AUSTRALIA
Division of Building Research,
CSIRO,
Graham Road,
Highett, Victoria 3190.

AUSTRIA
United Nations Industrial Development Organisation,
Vienna International Centre,
P.O. Box 400,
A-1400 Vienna.

BOTSWANA
Ministry of Local Government and Lands,
Private Bag 006,
Gaborone.

COLOMBIA
National Centre for Construction Studies,
Ciudad Universitaria C1145-Cra 30,
Edificio CINVA,
AA34219 Bogota.

EGYPT
General Organisation for Housing, Building and Planning Research,
P.O. Box 1170,
El-Tahreer Street, Dokky,
Cairo.

FRANCE
Centre Technique des Tuiles et Briques,
2, avenue Hoche,
75008 Paris.

International Union of Testing and Research Laboratories (RILEM),
12, rue Brancion,
75737 Paris

FEDERAL REPUBLIC OF GERMANY
Institut für Ziegelforschung, Essen e V, Am Zehnthof,
4300 Essen-Kray.

GHANA
Building and Road Research Institute,
University, P.O. Box 40,
Kumasi.

INDIA
Central Building Research Institute,
Roorke (Uttar Pradesh), 277672.

INDONESIA
Directorate of Building Research,
United Nations Regional Housing Centre,
P.O. Box 15,
84 Jalan Tamansari,
Bandung.

IRAQ
Building Research Centre,
P.O. Box 127,
Jadiriyah, Bagdad.

ISRAEL
Building Research Station,
Israel Institute of Technology,
Technion City, Haifa.

IVORY COAST

Société des Briqueteries de Côte d'Ivoire,
B.P. 10303,
Abidjan.

JORDAN

Building Materials Research Centre,
Royal Scientific Society,
P.O. Box 6945,
Amman.

MADAGASCAR

Centre National de l'Artisanat Malagasy,
B.P. 540,
Antananarivo.

MALAWI

Malawi Housing Corporation,
P.O. Box 414,
Blantyre.

NETHERLANDS

International Council for Building Research Studies
 and Documentation (CIB),
Weena 704, Post Box 20704,
3001 JA Rotterdam.

PAKISTAN

Pakistan Council for Scientific and Industrial Research,
Off University Road,
Karachi 39.

PAPUA NEW GUINEA

Department of Public Works,
P.O. Box 1108,
Boroko.

PHILIPPINES

Ceramic Association of the Philippines,
P.O. Box 499,
Makati, Rizal.

SUDAN

Building and Road Research Institute,
University of Khartoum,
P.O. Box 35,
Khartoum

SWITZERLAND

Technology and Employment Branch,
International Labour Office,
CH-1211 Geneva 22.

International Standard Organisation,
1, rue de Varembé,
CH-1211 Geneva

TANZANIA

Small Industries Development Organisation,
P.O. Box 2476,
Dar-es-Salaam.

THAILAND

Applied Scientific Research Corporation of Thailand
196 Phahonyothin Road,
Bankkhen,
Bangkok.

TRINIDAD

Caribbean Industrial Research Institute,
Post office box,
Tunapuna.

UNITED KINGDOM

British Ceramic Research Association,
Queens Road,
Penkhull,
Stoke-on-Trent ST4 7LG.

Building Research Establishment,
Bucknalls Lane,
Garston, Watford, Herts WD2 7JR.

Intermediate Technology Development Group,
9, King Street,
Covent Garden, London WC2E 8HN.

Intermediate Technology Workshop,
Corngreaves Trading Estate,
Overend Road,
Warley, West Midlands B64 7DD.

UNITED STATES

American Ceramic Society, Inc.,
4055 North High Street,
Columbus, Ohio 43214.

Volunteers in Technical Assistance,
1815 N. Lynn Street,
Suite 200,
P.O. Box 12438,
Arlington, Virginia 22209

UPPER VOLTA

Société Voltaïque de Briqueterie et de Céramique
B.P. 148,
Ouagadougou.

ZAMBIA

National Council for Scientific Research,
P.O. Box CH 158,
Chelston, Lusaka.

APPENDIX IV

LIST OF EQUIPMENT SUPPLIERS

                                            Type of equipment

AUSTRALIA
Automet Industries Pty Ltd.,   General equipment
P.O. Box 68,
88 Beattie Street,
Balmain NSW 2014,

BELGIUM
Sa Samic,
Hanswijvaart 21,
2800 Mechelen   General equipment

DENMARK
Niro Atomizer A/S,
Gladsaxevej 305,
DK-2860 Soeborg   General equipment

FRANCE
CERIC International,
18, rue Royale,
75008 Paris   General equipment

GHANA
Agricultural Engineers,
Accra   Trough mixer

INDIA
Raj Clay Products,
5 Mill Officers' Colony,
Ashram Road,
Navrangpura,
Ahmedabad 380009   Semi-mechanised equipment

ITALY

Unimorando Consortium,
Corson Don Minzoni 182,
14100 Asti — General equipment

KENYA

Christian Industrial Training Centre (CITC),
Meru Road, Pumwani
P.O. Box 729935
Nairobi — Crusher, table mould

NETHERLANDS

Joh's Aberson bv.,
8120 AA, Olst — Soft mud

UNITED KINGDOM

Craven Fawcett Ltd.,
P.O. Box 21, Dewsbury Road,
Wakefield, Yorkshire, WF2 9BD — General equipment

William Boulton Ltd,
Providence Engineering Works,
Burslem,
Stoke-on-Trent, Staffs, ST6 3BQ — General equipment

Croker Ltd.,
Runnings Road,
Cheltenham, Glos. — Pan mixer

British Ceramic Plant Manufacturers's
 Association,
P.O. Box 107,
Broadstone, Dorset BH18 8LQ — General information service

W.G. Cannon,
Broadway House,
The Broadway,
London SW19 — Fans

Auto Combustions Hoistrack Ltd,
Hartcourt,
Halesfield 13,
Telford, Salop. TF7 4QR   Oil burners

Intermediate Technology Workshops,
J.P.M. Marry and Assts. Ltd.,
Overend Road,   Crusher, table moulds,
Cradley Heath, West Midlands B64 7DD   Handling equipment.

Allied Insulators,
Albion works,
Uttoxeter Road,
Longton, Stoke-on-Trent ST3 1HP   Pyrometric cones and rings

Bair and Tatlock Ltd.,
Freshwater Road, Chadwell Heath, Essex   General laboratory equipment

William Boulton Ltd.,
Providence Engineering Works,
Burslem, Stoke-on-Trent, Staff ST6 3BQ   Clay machinery

Podmore and Sons Ltd., Shelton,
Stoke-on-Trent ST1 4PQ   Clay machinery

British Ceramics Service Co. Ltd.,
Bricesco House,
Park Avenue,
Wolstanton,
Newcastle-under-Lyme, Staffs, ST4 8AT   Kilns

Kilns and Furnaces Ltd.,
Keele Street,
Tunstall,
Stoke-on-Trent, Staffs, ST6 5AS   Kilns

Leonard Farnell and Co. Ltd.,
Station Road,
North Mymms,
Hatfield, Herts AL9 7SR   Testing apparatus and augers

UNITED STATES UNDERLINE_END

Interkiln Corporation of America,
P.O. Box 2048,
Houston, Texas 77252   General equipment

ZIMBABWE UNDERLINE_END

World Radio Systems,
Bush House, 72-72 Cameron Street,
P.O. Box 2772,
Harare   Crusher, table moulds

QUESTIONNAIRE

1. Full name..

2. Address..
   ....
   ....

3. Profession (check the appropriate case)

   Established brickmaker../__/
   If yes, indicate scale of production..................................

   Government official../__/
   If yes, specify position..

   Employee of a financial institution................................../__/
   If yes, specify position..

   University staff member../__/

   Staff member of a technology institution............................./__/
   If yes, indicate name of institution..................................
   ....

   Staff member of a training institution.............................../__/
   If yes, specify...
   ....

   Other, specify..
   ....

4. From where did you get a copy of this memorandum?
   Specify if obtained free or bought....................................
   ....

i

5. Did the memorandum help you achieve the following:
   (Check the appropriate case)

   Learn about brickmaking techniques you were not aware of /__/

   Obtain names of equipment suppliers /__/

   Estimate unit production costs for various scales of production/technologies /__/

   Order equipment for local manufacture /__/

   Improve your current production technique /__/

   Cut down operating costs /__/

   Improve the quality of produced bricks /__/

   Decide which scale of production/technology to adopt for a new brickmaking plant /__/

   If a Government employee, to formulate new measures and policies for the brickmaking industry /__/

   If an employee of a financial institution, to assess a request of a loan for the establishment of a brick-making plant /__/

   If a trainer in a training institution, to use the memorandum as a supplementary training material /__/

   If an international expert, to better advise counterparts on brickmaking technologies /__/

6. Is the memorandum detailed enough in terms of: Yes   No

   - Description of technical aspects..........................____.....____

   - Names of equipment suppliers.............................____.....____

- Costing information................................___.....___

- Information on socio-economic impact....................___.....___

- Bibliographical information............................___.....___

If some of the answers are 'No', please indicate why below or on a separate sheet:

....
....
....

7. How may this memorandum be improved if a second edition is to be published?..
....
....

8. Please send this questionnaire, duly completed to:

>   Technology and Employment Branch
>   International Labour Office
>   CH-1211 GENEVA 22 (Switzerland)

9. In case you need additional information on some of the issues covered by this memorandum, the ILO would do its best to provide the requested information.

www.ingramcontent.com/pod-product-compliance
Ingram Content Group UK Ltd.
Pitfield, Milton Keynes, MK11 3LW, UK
UKHW051128200426
11947UKWH00040B/1571